COMPUTING REALITY

Professor Dr. Masudul Alam Choudhury
Sultan Qaboos University
Muscat, Sultanate of Oman
&
Dr. Mohammad Shahadat Hossain
Chittagong University
Bangladesh

blue ocean press
tokyo

Published by:

blue ocean press, an Imprint of Aoishima Research Institute (ARI)
#807-36 Lions Plaza Ebisu
3-25-3 Higashi, Shibuya-ku
Tokyo, Japan 150-0011

mail@aoishima-research.com
URL: http://www.aoishima-research.com

ISBN: 978-4-902837-01-3 (Hardcover)
ISBN: 978-4-902837-13-7 (Softcover)

DEDICATION

To our colleagues, friends and students of the Postgraduate Program in Islamic Economics and Finance, Trisakti University, Jakarta, Indonesia.

ACKNOWLEDGMENT

We remain most grateful to our families, friends and many students who throughout the duration of this research project have contributed significantly in various ways. To our families we owe our gratitude for their patience at various times during this project. To our friends we owe their learned contributions and suggestions. To our postgraduate students of comparative epistemology and Islamic economics and finance at Trisakti University Jakarta, Indonesia, we owe the benefits of their incisive questions and critical outlook on various aspects of this project on scientific phenomenology.

We are also pleased with the acceptance of a refereed paper extracted from this book that will be presented in the Stockholm conference on Human Resource Costing and Accounting, Personnel Economics Institute, School of Business, Stockholm University, Sweden. Also, a significant outcome of this project is a planned Center for Advanced Study of Unified Systems (CASUS), Sultan Qaboos University, Muscat, Sultanate of Oman. Indeed, the scientific offshoots of this project are far many.

Importantly, we are pleased to have received an early publication agreement on this book from the Aoishima Research Institute, Tokyo, Japan. This inspired us to initiate an important research project that culminated into this book.

Indeed, we will be fully rewarded if the rare but deeply scientific message of this book on neurocybernetc phenomenology is assimilated by the scientific community at large and by the many decision-makers who run the public domain on human wellbeing and resource development. We hope to impart the contents of this book to our postgraduate students of comparative epistemology and computer modeling.

Professor Masudul Alam Choudhury
Department of Economics and Finance
College of Commerce and Economics
Sultan Qaboos University
Muscat, Sultanate of Oman

Dr. Mohammad Shahadat Hossain
Associate Professor and Chairman
Department of Computer Science
Chittagong University
Bangladesh

TABLES OF CONTENTS

FIGURES

CHAPTER 1: INTRODUCTION

Computing Reality is a rigorous study in the theory and application of the phenomenology of socio-scientific systems. This comprises machine (computer) logic and socioeconomic reasoning. The formal construction of the phenomenological perspective is applied to various socio-scientific issues and to the development of innovative computing logic. The totality of these two domains of concepts, formalism and application having both generality and particularity in it is what we will often refer to as neurocybernetic and system theory. The theme of phenomenology applied to computer logic and the formalism and application of a neurocybernetic and system methodology to socio-scientific issues means the presence of consciousness in machines (computational logic). This methodology of neurocybernetic and system formalism is logically derived and applied in a natural way to various social issues of epistemological re-construction of socio-scientific theory and its related computing logic for application.

Thus, machines (computer logic) and socio-scientific reasoning are studied in Computing Reality as mutually learning systems by way of interrelationships within and between them in methodological ways and by a unique and universal epistemological premise. Such a perspective of machine (computer logic) and socio-scientific reasoning makes the study of Computing Reality one of world-system consciousness. In such a phenomenological study, epistemological issues, logical formalism and applications are investigated together. Computing Reality thus opens up a new field of intellectual inquiry. It is a rigorous inquiry in the logic of unity of knowledge using neurocybernetic and socio-scientific systemic excitement of learning entities. Such system-entities are programmed and guided by moral and ethical consequences of the epistemology of unity of knowledge.

NEUROCYBERNETIC AND SYSTEM CONCEPTS
IN EPISTEMIC SENSE

The idea of neurocybernetic and systems that lies deep in the phenomenology of Computing Reality is sensitive to its particular epistemology that defines its scope within three phases of the complete phenomenological worldview. These three phases are summarized below and will be expanded at length in this work.

The meaning of neurocybernetic and systems emerges from the nature of the epistemological inquiry. In general, neurocybernetic and system theoretic studies convey ways of cognizing a nexus of entity-wide interrelationships pertaining to socio-scientific issues and computing logic that enable learning to appear and persist between them. Systems are formed by the nexus of such entity interrelationships with the goal of organizing and cognizing the relations in such ways that show them as being inter-causally dependent and to simulate the particular issues and problems toward attaining assigned functional objectives. Thus, while neurocybernetic is the study of the conceptual design of nexus of combinatorial interrelations between entities, system theory as defined by such nexus of neurocybernetic entities is goal focused. The word neurocybernetic is coined from the concept of the brain of computer logic and socio-scientific theorizing in the field of interdependent and unified entity relations.

To structure the neurocybernetic and system theory the epistemology of the nexus of entity interrelations and of the objective criterion for addressing it in the system theory must be precisely chosen. Epistemologies as theories of knowledge that emanate from differentiated, over-determined or mutational worldviews cannot work for unified neurocybernetic and system logic Such worldviews are found in neoclassical economic theory (Phelps, 1989), Marxist political economy (Resnick & Wolff, 1987), world-system theories (Wallerstein, 1974) in the literature and social Darwinism (Dawkins, 1976). Likewise, computer logic in neurocybernetic and system perspective of binary reasoning cannot reconstruct into systemic ethical thinking, conveying learning between its entities (Gleick, 1987).

In the end, the failure to derive and establish a unique and universal methodology governing ethical reasoning (by systemic learning) in machine logic and in socio-scientific theorizing, causes absence of a unified (learned) neurocybernetic and system theory in contemporary thought. The role of epistemology is therefore central in understanding new origins for construction of the phenomenology of neurocybernetic and systems reasoning and formalism pertaining to emergent issues and problems and explaining the evidential consequences (Sen, 1990; Choudhury, 2003a). Einstein commented that there is no science without epistemology (Bohr, 1985).

Formalism of the phenomenology of socio-scientific thinking along with the problems of machine (computing) logical theory emanate from background epistemologies. A particular epistemology is formally cognized at the stage of ontology. This is the formal phase for the development of the being and becoming of the formal neurocybernetic theory in the light of the foundational epistemology. Epistemology and ontology together then lead the whole system of computer logic and its socio-scientific theorizing into concrete applications. This is the ontic or evidential phase of the phenomenology of computer logic and socio-scientific theory. Both of these are shown to be premised uniquely and universally on the same worldview of the underlying epistemology. The complete phenomenological construction for computer logic in concert with socio-scientific reasoning is thus made up of the three phases of a comprehensive theory and application.

THREE PHASES COMPRISING THE PHENOMENOLOGY
OF UNITY OF KNOWLEDGE

The three phases comprise firstly, discovery of the fundamental Epistemology. Secondly, the abiding premise of the Epistemology helps in cognizing the underlying theory pertaining to the understanding, formalism and analytical construction of a unique and universal worldview for addressing any and all issues and problems. This is the phase of ontology emanating from the foundational epistemology. It formalizes the emergent thought related to particular issues and problems within the construction of the unique and universal worldview. Thirdly, the phases of foundational epistemology and ontology together combine to define, explain and organize the world-systems of evidences. At this stage combining epistemology and ontology together brings about the application of the emergent methodology to specific issues and problems. This takes the form of evaluating unifying interrelations between socio-scientific variables (system-entities), while casting such analytical evaluations in the framework of neurocybernetic and system reasoning.

The latter too are derived by reference to the epistemological and ontological precedence. State and control variables, institutional matters, social structures, policies, preferences and prescriptive guidance, all emanate at the third phase in concert with the first two phases. The total phenomenological methodology of learning systems in unity of knowledge is thus completed. Such learning experiences in unified world-systems are incessantly evolutionary over similar processes. The phenomenological worldview of unity of knowledge over the three phases will be deeply studied in this work.

4

PHENOMENOLOGY OF UNITY OF KNOWLEDGE

The phenomenology of unity of knowledge will be premised on its specific epistemology of the unique worldview. Such an episteme will be shown to logically establish pervasively complementary interrelations between systems and their entities in respect to the specified goal of attaining higher levels of wellbeing. The principle of pervasive complementarities will be explained in a substantive methodological sense within the purview of the intra- and inter-systemic interdependent relationships between the entities in the perspective of the three properties of such relational behavior of systems.

These properties are firstly, pervasive interaction by virtue of diversity of interpreting the world-systems in the framework of discursive phenomenon. This also conveys the meaning of symbiosis between cognitive and systemic entities as the sign of learning. Systemic interactions in the discursive environment carry with them the ontology derived from the epistemological core for cognizing the world-systems according to such precepts, rules and instruments that explain the symbiotic nature of all world-systems and their entities in respect to specific issues and problems under investigation.

The second property is the convergence into consensual regimes of choices and decisions following the discursive experience crossing diversity and possibilities. In this way organic unity of learning by interaction and convergence to consensus is established. We refer to this knowledge-induced experience and learning as process, opposed to determinism. Thus Interaction leading to Integration with the evaluation of consequences of learning at the end and leading thereafter into further co-Evolution of the same learning experience will be sometimes referred as the interactive, integrative and evolutionary process (IIE).

The stage of convergence of interactions into consensus of choice and preferences in the discursive environment will be referred to as the Integrative stage. The use of the Interactive and Integrative stages of inter-entity learning in

5

neurocybernetic and systemic correspondences is thus implicative of the general phenomenon of learning by the relational concept of unity of knowledge. This is intrinsic both in cognitive entities, including the ontology of formalizing the learning relations as derived from the foundational epistemology, and in computing logic. The experience in unity of knowledge as in the interactive-integrative process is also endowed to institutions, laws, rules, policies and instruments. These various facets of the phenomenological model will be shown to induce the interactive-integrative learning experience.

The third property of learning systems that we will develop along with the Integrative-Integrative neurocybernetic and system theory in generalized framework of all world-systems is that of co-evolution of the same experience of learning systems over phases of learning. In this way, the learning experience of interaction leading to integration appears over several problem dimensions. These are firstly, the continuously occurring knowledge-flows generated in the interactive and integrative stages. This process stage is co-evolved over phases of learning by interaction and integration and their consequential evaluation in reference to unity of knowledge as the episteme. Secondly, the observation of the knowledge-flows and of the knowledge-induced system-entities is observed over time, but not caused by time. Thirdly, the continuity of the neurocybernetic systemic processes extends over spatial and cognitive domains at every point in time pertaining to particular issues and problems and generalized theory.

We will close the unique and universal phenomenological model of the learning universe in its systemic entirety by means of the three properties of every process of learning, namely Interaction, Integration and Evolution (IIE-process). The methodology of unity of knowledge as the foundational epistemology of the complete phenomenological model will thus be established methodologically and explained textually in this work. The resulting worldview of unity of knowledge will form the complete reasoning of the neurocybernetic and system theory. Such a study will be taken up at the level of computing logic and socio-scientific system theory.

THE CENTRAL EPISTEMOLOGY OF UNITY OF KNOWLEDGE

Unity of knowledge as the episteme of Computing Reality towards formalizing the building blocks of a unique and universal neurocybernetic and system theory of computing logic and socio-scientific reasoning will be central in our methodological formulation. From the above mentioning, the comprehensive neurocybernetic and system theory premised on the epistemology of unity of knowledge comprises three stages.

Firstly, the epistemological foundation is appropriately determined. This goes through the process of search and discovery over a substantive survey of the literature in which classical and contemporary scholarly contributions to this field are examined. From the substantive search discovery and re-search we will ineluctably converge into the logical premise of unity of all laws in the primacy of the divine laws. This search and discovery is neither of a religious orientation nor nihilistic. It is methodologically determined by mathematical reasoning on Universality. The mathematical derivation will be left to the appendix as will be various similar technical derivations. The understanding of such technical tracts will be explained in the text. Determination of the uniqueness and universality of the epistemological premise of unity of knowledge, further specified as the critical building block of the corresponding type of neurocybernetic system theory, will thus be the first analytical pursuit of this work.

Secondly, we will show how with methodological power of analytical argumentation the same epistemology yields the nature of the ontology for cognizing the formal framework of analysis, the design of reasoning in computing logic extended to socio-scientific reasoning. The general methodology and theory in the context of the epistemology and ontology of learning systems will be

7

particularized to specific problems of computer logic and socio-scientific categories. The questions asked against the background are these. Can any other epistemological and ontological perspective lead to the same kind of learning neurocybernetic system design? Can a unique and universal theory of such neurocybernetic systems be found in any epistemology and ontology other than the divine laws, which in a logical sense unifies all laws, rules and cognitions? We will quest for and confirm the answer in the uniqueness and universality of the divine laws as the premise.

The meaning of ontology in our understanding of system logic (Gruber, 1993) is to cognize the methodology, formalism and design of systems according to the epistemology of unity of knowledge in learning systems. We will argue that such learning systems, as embedded ones, form the true nature of the world-systems, as opposed to differentiated systems that promote and explain self-interest, conflict and competition between system-entities. We will prove that only the universal methodology emanating from the epistemology and ontology of unity of knowledge can answer both the perspectives – of learning (unity) and de-learning (de-knowledge).

In the third stage, the phenomenological model as mentioned earlier will be completed by inducing the epistemology and ontology of unity of knowledge into concrete problem solving. This is the empirical stage and result from and into inferences, policies and predictive results. Several complex issues, such as of model formulation, relational epistemology in such model formulation, preference formation, institutionalism, special quantitative methods emanating from the treatment of the phenomenological perspective of unity of knowledge, the nature of the knowledge-induced universe and domains will be studied by complex reasoning. Complexity is meant here as embracing the richness of the issues and problems in the neurocybernetic learning world-systems of all types.

The stage of evidential categories that finally results are referred to as the ontic stage of the total three-phases of the complete model of phenomenology in unity of knowledge for learning systems. The ontic and ontology are different parts of systemic reasoning, in that ontology of cognitive reasoning and formalism premised on the methodology that ensues from the primal epistemology.

Yet even at the ontological stage of systemic reasoning the identification of specific variables, rules, relations, instruments etc. can be included for designing the comprehensive model that can be ontologically cognized. In the ontological stage there is no need to actually apply the model conceptualized as logical formulation in reference to the epistemology of unity of knowledge. Ontology thus presents a deductive reasoning for its inductive application at the ontic stage. It is at the ontological stage that the above-mentioned IIE-process will be formalized in terms of the properties of all neurocybernetic learning systems.

On the other hand, the ontic stage quantifies and applies the formal model to specific issues and problems so as to turn up results by reference to the conceptual model that has been established at the ontological stage in reference to the epistemology of unity of knowledge. Quantification and application of the model of the phenomenology of unity of knowledge to particular issues and problems will bring out many specialized concepts and methods that need to be designed for simulating the formal model (Boland, 1991).

THE EPISTEMOLOGY OF DIVINE ONENESS IN THE PHENOMENOLOGICAL MODEL

The epistemological origin of divine oneness transported into neurocybernetic and system theory by the force of analytical methodology is found to be deeply moral and ethical in nature. The nature of morality and the derived ethics in the systemic sense of neurocybernetic and socio-scientific reasoning presents formalism in endogenous ethical relations among all system-entities. An example is that of understanding capital and capital accumulation (Zohar & Marshall, 2004; Choudhury, 2004), and thereby, all kinds of market exchange as organic processes (Holton, 1992), all being induced by systemic ethics emanating from the epistemological and ontological premises.

The theme of endogenizing ethics so as to capacitate learning in the system with unity of knowledge will receive special attention in this book when studying economic, financial and social systems and the neuroscience of the supercomputer (Humaine, 16 Dec. 2003). Indeed we will argue that the supercomputer that confronts complex issues and problems must decipher the logic of simulation of such complex systems by the methodology of the Interactive, Integrative and Evolutionary Process (IIE-process) within the complete phenomenological worldview.

The learning character of system-entities according to the nature of the issues and problems of learning world-systems results in complex problems. Yet the phenomenological model is simple in its structure. But its extensions to multidimensional learning and interconnecting systems, their system-entities and relations will prove to be complex due to the richness of such neurocybernetic learning systems.

Computer logic in such nexus of interrelated systems is thus inherently linked to socio-scientific reasoning in a methodological way (Shakun, 1988). The complexity is not in the phenomenological structure. Rather, complexity is caused by extensions of the universal phenomenological model of unity of knowledge to multidimensional fields and objectives. This property of the complete phenomenological model will be shown to make computer logic and socio-scientific reasoning, and thereby, the ontological and ontic formalism to be of probabilistic types. We will show how probabilistic reasoning enters predictive models derived by the phenomenological approach. Such probabilistic ways of systemic reasoning and quantitative analysis will be compared with similar methods in Quantum Physics.

FURTHER CONCEPTS IN THE PHENOMENOLOGICAL MODEL SUMMARIZED

A number of other critical concepts linked to the phenomenological methodology and quantitative model of unity of knowledge need to be introduced now. These are 'circular causation' governed by the 'principle of pervasive complementarities' (Choudhury, 2003b), 'social wellbeing' as the objective criterion quantifying attained levels of unity of knowledge pertaining to the issues and problems under study, and 'system sustainability'.

Circular Causation

Circular causation is represented by reflexive *inter*relationships between the component variables of the objective criterion called social wellbeing in the process of solving the simulation problem of social wellbeing function. The variables, and hence their relations, are induced by ontological knowledge-flows and the accompanying rules and guidance arising from the epistemology of unity of knowledge. Circular causation is specifically represented by recursively relating any particular variable to rest of the variables that all together as system-entities, enter the social wellbeing function as the simulated objective criterion.

Simulation is invoked as opposed to optimization method (Choudhury & Korvin, 2002) in order to indicate the evolutionary learning feature of the IIE-processes. The IIE-processes underlie the methodology of the total phenomenological model in the light of the epistemological, ontological and ontic stages of neurocybernetic learning systems. By virtue of the circularly recursive interrelations between the system-entities, circular causation represents the unification process between these variables. Circular causation involving recursively generating the knowledge-flows premised on unity of knowledge

11

induces the system-entities by knowledge-flows in the underlying IIE-discursive processes. Such processes characterize the entire phenomenological model of unity of knowledge as systemic learning proceeds on. Such a learning process firstly determines the understanding and choices of rules, guidance, preference formation, variables and instruments that are ontologically derived in the IIE-discursive process in reference to the epistemological premise, and that subsequently lead into quantitative simulation to yield pertinent results for social decisions, social futures, actions and responses.

Complexity

Issues of complexity in measurement, estimation, model selection, and probabilistic perspectives of knowledge-induced fields of variables of the social wellbeing function are all actively pursued. We will have occasion to cover these themes in reference to logical reasoning, as invoked by the kind of neurocybernetic and system theory to be formalized, spanning its conceptual and applied levels. In the complexity perspective, the Quantum theoretic nature of the probabilistic limit points will be highlighted in order to establish the existence and continuity of evolutionary equilibriums that characterize the evolutionary nature of IIE-learning systems (Grandmont, 1989; Choudhury, 1993). The complexity phenomenon is explained, firstly in a textual nature in the body of the relevant chapters for the benefit of the common reader. Technical matters are left to the appendixes for the benefit of the technically prepared reader. The existence and pervasiveness of evolutionary equilibriums in learning systems will be found to characterize most issues and problems at their various levels of learning along the IIE-process worldview of unity of knowledge taken over the epistemological, ontological and ontic stages of the complete phenomenological model of unity of knowledge.

12

Wellbeing Objective Criterion

Next we need to introduce at this time the idea of wellbeing as the criterion function of the simulation problem of recursive learning in unity of knowledge. Wellbeing is the unique objective function for addressing every issue and problem by means of the general model of phenomenology of unity of knowledge in all types of neurocybernetic and system theory, computer logic and socio-scientific reasoning. The inception of this model arises at the epistemological and ontological stages but it takes quantitative form for an engineering treatment of the IIE-process at the ontic stage.

The wellbeing function is the criterion for measuring the degree to which unity of knowledge is realized in the learning system for investigating specific issues and problems. Its functional form is a technical exercise in model selection. For our case we will always take the model selection for the wellbeing function in the product form in knowledge-induced variables pertaining to the issues and problems under investigation. The product form of the wellbeing function in the selected variables implies codetermination of the variables in the total systems sense. Such variables appearing in their complex but pervasively complementary fields as organic unity caused by the learning process, can be trans-systems or within given systems, issues and problems under study.

The total model of wellbeing will be treated as a simulation problem of the wellbeing function in terms of the inter-linkages between the variables, which will be explained by means of recursive feedback between them, yielding positive signs of complementarities. The simulation will be carried over knowledge-flows affecting the variables, and thus enabling them to complement with each other. Pervasive complementarities between system-entities are the manifest signs of unity of knowledge in the framework of circular causation in the simulation problem. Such simulation conveys the learning process between system-entities, for any particular problem under investigation.

Hence the variables form categories comprising firstly, systemic variables (also called state variables); secondly, preference variable that carry knowledge-flows; and thirdly, policy instruments. The second and third types of variables are referred to as control variables. The additional relation in the overall simulation relations is the recursively determined knowledge-flow variable, which affects all variables in the evolutionary motion of circular causation. In the probabilistic limiting sense the continuous assignment of this variable in evolutionary IIE-processes is necessary to enable the whole system to learn and evolve. The learning by the IIE-process dynamics causes both state and control variables to become endogenous in nature. This will be treated as a major contribution to the understanding of endogenous preferences, institutions, policies and instruments. This contribution is contrary to the exogenous treatment of such variables in received economic, financial and social theories.

System Sustainability

An important focus of this work will be on developing the functional idea of system sustainability as the powerful premise for understanding, establishing and protracting the essence of unity of relations in continuity and continuums of systems. In the learning neurocybernetic of unity of knowledge the idea of system sustainability is extended consequentially to the continuity and pervasiveness of system-entity reflexive *inter*relations both intra- and inter-systems.

As mentioned earlier, our research in this work for the ultimate origin and permanence of unity of knowledge will lead us into the acceptance of the divine laws as the 'originary' episteme (Howard, 1985; Foucault, 1972). The entire design, explanation and formalism of the model of phenomenology of unity of knowledge will be premised on the epistemology of the divine laws as the carrier of unity of knowledge in everything. The logical derivations of concepts, rules, guidance, attenuating variables and their relations in the simulation of the wellbeing function will all be conditioned by the knowledge-flows that emanate from the finality of the divine episteme as the explainable and applicable premise of the general system. This will form the worldview of the unique and universal model of unity of knowledge.

System sustainability is thereby understood within this kind of epistemological context taken through the ontological development into ontic application and thereafter co-evolved. We will understand by the concept of system sustainability the self-organizing continuing nexus of interrelationships according to the episteme of unity of divine knowledge in the phenomenological model of unity of knowledge.

OVERVIEW OF THE BOOK

Chapter 2 reviews and presents alternatives to the received literature in the area of neurocybernetic and system paradigms. Its objective is to make the reader aware of the need to establish the epistemology of unity of knowledge, where this is presently non-existent in mainstream system understanding of socio-scientific reasoning.

Chapter 3 develops a computerized model of computing reality called the MCR and applies it to controlling flood-risk in Bangladesh. Various elements of the process model of socio-scientific phenomenology in respect of its relational epistemology, ontology and ontic stages of formalism are built into the MCR. The central role of circular causation between the variables of the flood-risk control problem is explained to show how efficient decision-making on flood-risk control can be realized.

Chapter 4 further extends the neurocybernetic and system theory of relational learning to the case of social sciences. It differentiates between the emanating complex analyses in non-linear spaces against the linear analysis of differentiated world-system studies. Within this purview of complexity versus

•

linearity contrasting doctrines of the embedded systems versus differentiated systems are discussed. In the end, it is concluded that methodological individualism remains the permanent feature of received mainstream economic and social theory.

Chapter 5 deals with the origin of rationalism and the ideas relating to this among several occidental philosophers. The problematique of continuity of knowledge over space and time into the domains of abstraction and the socio-scientific world-systems is pointed out in such rationalist philosophies. Thus the project of neurocybernetics and system theory in unity of knowledge is shown not to be realized in the rationalist framework. Issues of institutionalism and moral philosophy as an area in neurocybernetics and system theory are studied in the light of rationalist philosophy against unity of knowledge.

Chapter 6 deals with yet another rarely recognized but important theme of neurocybernetic and system theory. This comprises a study in the realm of moral and ethical laws. The chapter addresses the theme of how human preferences can be transformed by this kind of consciousness. The intellectual thought and contributions of the classical Islamic scholar, Abu Hamid Al-Ghazali is the center of this neurocybernetic and system study in comparative perspectives, showing how this classical moral treatment is still fresh in the world of highly analytical thinking and application.

Chapter 7 is a formalization of the methodology of unity of knowledge. The foundation of such a model is shown to be in the Qur'an according to the divine laws of oneness of God. Several Qur'anic ways of explaining the unique and universal methodology of unity of the divine laws are explained for the socio-scientific order and details in this respect.

The text ends with a conclusion of the important themes of this book. The rarity of this book and the underlying ideas in the realm of original research is highlighted. Further directions of research and critical thinking are pointed out for the benefit of the deep researchers in the area of computing reality

THE CAPTURING AUDIENCE OF THIS BOOK

This book is an original contribution in establishing the building blocks of a neurocybernetic and system theory for socio-scientific reasoning and computation premised on unity of knowledge. In this area, the background research that led to the inception of this book is a search for the ultimate source of unity of knowledge as the episteme. The research and this resulting book, thereby present rigorous formalism and application pertaining to socio-scientific issues and problems. These issues and problems will come from the field of computer logic and computer design, economics, finance and sociology.

The style of the book, being divided between textual readability and rigorous treatment of socio-scientific concepts in the field of neurocybernetic and system theoretic treatment of unity of knowledge, will appeal to all categories of readers. Non-technical readers will find the material accessible in understanding emergent new ideas in the realm of system sustainability explained by the logic and reasoning of computing reality. Technically prepared readers will find the whole material provoking a new scientific realm of thinking and its modus operandi of application to specific issues and problems. Many of the mainstream methodologies and methods of formalizing and analyzing problems will be questioned and alternatives provided within the phenomenological model of unity of knowledge.

The book is thereby expected to provide serious research reference materials to students, researchers and scholars of philosophy of science, economics and epistemic, computer scientists and sociological schools. Likewise, it should be accessible and useful for the informed readers in their search for new perspectives in the world of learning and intellectual thought. It is possible that a

17

number of postgraduate thesis and research topics will be derived from this work. Thereby, the ideas presented here in their original perspectives can be extended to a wider audience as intensification of interest in the area continues. Most university and national libraries around the world should acquire the book for the scholarly and educative purposes for which it is meant.

REFERENCES

Bohr, N. 1985. "Discussions with Einstein on epistemological issues", in H. Folse, *The Philosophy of Niels Bohr: The Framework of Complementarity*, Amsterdam, The Netherlands: North Holland Physics Publishing.

Boland, L.A. 1991. "On the methodology of economic model building", in his *The Methodology of Economic Model Building*, pp. 39-63, London, Eng: Routledge.

Choudhury, M.A. 1993. *Unicity Precept and the Socio-Scientific Order*, Lanham, Maryland: University Press of America.

Choudhury, M.A. 2003a. "A Qur'anic theory of deontological consequentialism", in his *Explaining the Qur'an Book 1*, Lewiston, New York: The Edwin Mellen Press, pp. 345-396.

Choudhury, M.A. 2003b. "Islamic political economy", *Review of Islamic Economics*, No. 13, pp. 49-72.

Choudhury, M.A. 2004. "A monetary system with 100-percent reserve requirement and the gold standard", in M.A. Choudhury & M. Z. Hoque, *An Advanced Exposition of Islamic Economics and Finance*, Lewiston, NY: The Edwin Mellen Press, pp. 187-198.

Choudhury, M.A. and Korvin, G. 2002. "Simulation versus optimization in knowledge-induced fields", *Kybernetes: International Journal of Systems and Cybernetics*, 31:1, pp. 44-60.

Dawkins, R. 1976. *The Selfish Gene*, Oxford, Eng: Oxford University Press.

Foucault, M. trans. A.M. Sheridan 1972. *The Archeology of Knowledge and the Discourse on Language*, New York: Harper Torchbooks.

Gleick, J. 1987. *Chaos, Making a New Science*, New York, NY: Viking Penguin.

Gruber, T.R. 1993. "A translation approach to portable ontologies", *Knowledge Acquisition*, Vol. 5, No. 2, 199-200.

Holton, R.J. 1992. *Economy and Society*, London, Eng: Routledge.

Howard, D. 1985. *From Marx to Kant*, Albany, NY: State University of New York Press.

Humaine, 16 Dec. 2003. *(http://emotion-research.net/Members/KCL)*:

Phelps, E. 1989. "Distributive justice", in Eatwell, J. Milgate, M. & Newman, P. eds. *The New Palgrave, Social Economics*, New York, NY: W.W. Norton, pp. 31-34.

Resnick, S.A. & Wolff, R.D. 1987. *Knowledge and Class, A Marxian Critique of Political Economy*, Chicago, ILL: The University of Chicago Press.

Sen, A. "Economic judgements and moral philosophy", in *On Ethics and Economics*, pp. 29-34, Oxford, Eng: Basil Blackwell Ltd. 1990.

Shakun, M.F. 1988. *Evolutionary Systems Design, Policy Making under Complexity and Group Decision Support Systems*, Oakland, CA: Holden-Day, Inc.

Wallerstein, I. 1974. *The Modern World Systems*, New York, NY: Academic Press

Zohar, D. & Marshall, I. 2004. *Spiritual Capital*, San Francisco, CA: Berrett-Koehler Publishers, Inc.

CHAPTER 2: REVIEW OF THE LITERATURE: CYBERNETICS IN SOCIO-SCIENTIFIC SYSTEMS

Recent contributions to the theory of systems and cybernetics have opened serious thinking in the direction of 'globally' complementary and embedded systems (Xuemoue & Dinghe, 1999; Yolles, 1998; Johannessen, 1998; Hubner, 1983). Also in the theory of management systems there is fresh inquiry on the direction of learning cybernetics of management (Jackson, 1993). We (Choudhury, 1999a; Hossain, Choudhury & Mohiuddin, 1998) have contributed to this growing field of socio-scientific theorizing of systems and applied formalism to the case of Bangladesh on the theme of causes and controls of floods. In earlier times there were the masterful contributions by Herbert Simon (1957) on models of man as artificial intelligence in organizational behavior.

EXAMINING THE PRINCIPLE OF COMPLEMENTARITIES

The central focus of the above-mentioned contributions is on the following directions: All analytical forms of physical or human learning systems imply a continuous regime of inter-connectedness among the entities of such systems. This form of complementarities is a sign of neural learning out of systemically generated knowledge emanating from a given epistemological outlook. Such an outlook of theory is endowed to particular cybernetic ways of viewing the complementary world-systems. The principle of pervasive complementarities invoked in this work conveys the picture of ceaselessly and continuously interconnected nexus of world-systems that learn as embedded sub-systems in relation to each other. Such sub-systems unify between themselves in terms of the relationships generated in and between them by the variables and agencies that characterize the sub-systems. The focus of all sub-systems of 'globally' extensive world-system is thereby on integrated learning arising out of interaction between the systemic entities. Such a systemic experience takes place in a dynamic sense of creative evolution as the process of interrelationships continue to advance across overarching systems.

The combination of interactively generated integration across evolutionary dynamics of interrelations, that is co-evolution of processes, is the meaning of systemic learning. Complementarities are thus the result of universalizing the systemic learning process of overarching embedded and interconnected sub-systems of the world-system. In this work we will spend a considerable length in formalizing and applying such 'universally', that is 'globally' complementary, and thereby learning systems that are unified by the phenomenology of interactively generated integration and creative evolution.

The word phenomenology itself will acquire a technical meaning as we proceed on in subsequent chapters. Briefly speaking, phenomenology means the degree of consciousness gained by the learning processes involving entity interaction followed by their convergence into consensus or equilibriums, thereafter evolving further by co-evolution of similar processes of learning. From the extension of complementarities across overarching systems and their embedded entities as variables, agencies and their relations, will be derived the technical meaning of the unified nexus.

We can therefore infer even from this first stage of definitions and explanations of certain terms that the conception of pervasive complementarities forms a principle that overarches systems. The result is a form of circular causality. This means continuous learning by feedback between the entity-relations of the learning systems. The overarching nexus of complementarities form the continuum. The overall picture of scientific formalism now is to view the world-systems as being unified by learning by the unique and universal law of unity of knowledge as the binding epistemology.

SPECIFICS ON THE STUDY OF COMPLEMENTARISM
IN THE REVIEW OF THE LITERATURE

Organizational Theory as Learning System

Herbert Simon (op cit, 1960) viewed the models of man as an artificial intelligence of organizational behavior. Artificial intelligence is a means of automatically generating a learning process based on initially given descriptions and tasks for performing in the systemic relations between the entities. The epistemic premise for Herbert Simon is the capacity to rationally compete, choose and organize firm (organizational) behavior through complex learning processes. There are two dimensions of the learning perspective.

Firstly, complexity of organizational forms leads to compartmentalized decision-making on the basis of the sub-systemic preferences, rationality and information stock. The result is that such competitive and rationally developed decision-making in complex systems breaks up the organization into hierarchies and compartments. Simon's organizational forms are essentially hierarchical in nature across the complex structure of organizational behavior.

Secondly, Herbert Simon's organizational behavior theory in the perspective of artificial intelligence hierarchically organized and extended in complex systems is like Darwinian mutation with an increasingly weak link between evolutionary hierarchies. The kind of relationship that finally prevails in this kind of mutation dynamics within and across complex systems conveys the persistence of competition and rational choice caused by hegemony for control of information to make simulated optimal decisions within each compartment of the evolutionary hierarchy. Simon's organizational system is thus caused by learning within sub-systemic entities, which subsequently results in mutation into evolving hierarchies. Such hierarchies are mutated ones. The hierarchies between them become decentralized and are independently capacitated to make decisions in a rational process of its own. Yet Simon wants to limit the degree to which such

23

hierarchical decentralization should proceed. Consequently, even in the mutation system configured by Simon there is a prospect of weakly interconnecting between sub-goals within the organizational totality.

The organizational rationality, referred to by Simon as 'bounded rationality' because of information asymmetry and unequal levels of the information stock owned by competing firms, can be explained by the scheme given below. This is a simplification of Simon's O-Theory (Simon, 1952-53, 1957).

Denote organizational Relations in Simon's perspective of the economics firm by R_i, $i = 1, 2, 3, 4, 5$ with,

$R_1: I \to C$

I denotes Interaction within a hierarchy of decision making, and hence in levels of decentralization.

C denotes Cohesion between preferences of entities (agents) in given hierarchies.

$R_2: C \to D$,

D denotes Diversity in complex forms.

$R_3: D \to C$

$R_4: C \to I$

$R_5: I \to D$.

The feedback between these relations is broken by the missing $R_6: D \to I$. If the transitivity condition is applied between R_3 and R_4, this would yield R_6. But the question remains whether in complex organizational decision making the postulate of transitivity can be applied just to preserve the axiom of economic rationality. The breakdown of the axiom of transitivity was pointed by Condorcet & Caritat (1785) and lately by Sen (1977). In social choices as an example of complex decision making, also emulated by the preferences and menus of organizational theory of the firm, transitivity postulate is satisfied only under conditions of perfect competition in the core of economy (Debreu, 1963). Alternatively, a dictatorial condition must exist to limit the discursive process and bring about agreement (Arrow, 1951).

An example of such anomaly in social decision making is found in political gallop polls. Predictions of polls (based on rational behavior) although transitively inferred by repetition prior to election can be contradicted by actual political results. This proved to be the case with the Spanish and Indian democratic elections in recent times, despite their transitively predicted popular predictions on the election result. In the case of the Egyptian election re-electing President Hosni Mubarak of Egypt, transitively established election result proved to be the case of a rigged presidential election.

Likewise, approval of project feasibility study by most departments in an organizational hierarchy although transitively predicted can be rejected by some departments, thereby blocking the approval of the feasibility report. Take the example of transitively expected ratification of the Kyoto Agreement on pollution control for the environment by the countries and members of the Kyoto Summit on environment. Yet the treaty was vetoed by the U.S.A. Government. Similar was the case with the bio-diversity program of the Rio Summit despite its overwhelming transitively expected ratification by most members of the United Nations who were present in the Summit. Yet the U.S. Government opposed it. Another example is of the Big Carrot participatory industrial organic farm products supplier in Toronto (Morgan & Quarter, 1991). Despite transitively expected decisions to establish participatory enterprise, Big Carrot has recently given rights and privileges of selective decision making to hierarchical groups in the corporation and its financing sources.

NEUROCYBERNETIC THEORY OF MANAGEMENT SYSTEMS

Another perspective of system theory of organizational behavior arises from management theory. Management theory deals with the method and art of organizational governance. It need not be driven by a commitment to abide by a given epistemology of the background organizational theory. Also in diverse social systems different management theories or methods of organizational governance can abide. One can think of some extreme cases.

Weber's criticism of modern development in organizational theory *qua* management methods was meant as a herald to the coming age of individualism in which capitalism bureaucracy and rationalization of the governance method would prevail supreme. This would kill the values of the individual in which he self-actualizes with the collective. Mommsen (1989, p. 111) writes on Weber's concern with the future development of bureaucracy and rationalization in organizational theory enforced by the power of management methods. Weber fears that this gaining hegemony would petrify the liberal idea: "... Weber was all too aware of the fact that bureaucratization and rationalization were about to undermine the liberal society of his own age. They were working towards the destruction of the very social premises on which individualist conduct was dependent. They heralded a new, bureaucratized and collectivist society in which the individual was reduced to utter powerlessness." Weber was thus lost between the dilemma of pure individualism of liberal making (Minogue, 1963) and collectivism led by self-seeking individualism formed into governance. This Weber feared would destroy the fabric of the value of individualism on which liberalism was surmised to have been erected.

Today the International Monetary Fund and its sister organizations such as the World Trade Organization takes a global governance view based on Weber's kind of opposed liberal perspectives. Firstly, the IMF (1995) promotes global ethics under the guise of rational choices and human consensus of nations. On the

26

other hand contrarily, the IMF and its various sister organizations impose conditionalities for stand-by funds to developing countries. In doing so, the Bretton Woods Institutions maintain strict abidance with macroeconomic policies and designs. The IMF policies and conditionalities together with the World Bank's structural adjustment as development management practices have brought about failed futures for many countries (Singer & Ansari, 1988).

The management perspectives prevailing in international development organizations are a prototype of the preferences of self-interest and methodological individualism transported to organizational behavior and enforced by global governance as an extreme form of global management. One can refer to the public choice theoretic nature of such organizational preferences and management behavior explained by Ansari (1986). Many examples of this kind of international control and governance within global capitalism are prevalent in institutions such as the WTO, Basle II and the regional development organizations. The latter are forced to pursue the same directions as the international development financing institutions by design and interest (Emmerij, 2000). The transnational corporations too become engines for managing capitalist globalization in this order (Sklair, 2002).

Jackson (1993) sees management systems to be a designing of social reality as perceived by the principal-agent game within organization rather than being led by any kind of epistemological premise. Consequently, social reality is constructed in management system theory as a perception of the agents. This allows for the contest of individual wills served by those who mould these preferences in organizations. The institutional mediums either propose or enforce the preferences of methodological individualism in society at large.

In the neurocybernetic concept of organizational management theory that Jackson proposes, it can be inferred that such a systems perspective of governance serves only to deepen the methodological individualism and competition and contest of wills that ensue from management practices. Management system theory is thereby not necessarily premised on a learning behavior with unity of knowledge for attaining a common goal of mutually perceived social reality. Yet a learning practice in unitary management systems remains a possibility

Jackson sees three kinds of management systems reflecting social reality as it is perceived within the governance circle. Yet in the absence of a well defined epistemological premise and a morally determined social goal for all, none of these types of management systems can pave the way to human wellbeing intertemporally, established by sustainability and a shared vision (Inglott, 1990). The reason for this human predicament of management theory is that a management system can be very integrative and consensual and yet utterly hegemonic on common wellbeing.

In world political economy, the North has established an effective arsenal of co-operative development financing institutions but at the detriment of wellbeing for the poor South. This is most pronounced on matters of collective military pacts, belligerence and institutional and technological monopoly of the North on the South. Likewise, an integrative management of governance system can be enforced by the dominant force. The abuse of United Nations authority by the United States, Britain and their alliance on matters of war and peace proved to be true in the case of invasion of Iraq in Gulf War II.

A *coercive system* is one that is purely of the individualistic type. Many transnational corporation management practices in capitalist globalization and the political management of war by force can be categorized as coercive systems. Coercive management systems are the principal ones in today's global governance. It is also this kind of efficient governance by force that was presented as a model of dominance and national control by Machiavelli (1966).Cummings (2005) gives an incisive coverage of this kind of hegemonic management of global governance in many areas of capitalist globalization in present times.

Other forms of management systems pointed out by Jackson are the *pluralistic* and *unitary* types. *Pluralistic* management is a principal-agent game in which the interest of stakeholders is attained by consensus, despite the existence of diverse and opposing views on the issue under discourse, but being guided by the willingness of participants to coordinate and co-operate. An example of this case is industrial democracy, where management and workers can arrive at consensus on management issues despite their opposing views on particular issues. We have already seen that even in the pluralistic model of management for governance within the Bretton Woods Institutions, self-interest and power-centric approaches are entrenched in the hands of the industrialized nations over the developing ones.

The *unitary* model of management systems is based on pre-existing agreement among participants on assigned goals and rules in institutional discourse over issues. The abidance by liberalism as the foundation of the cultural make up in western democracy and its social reality is an example that prevails over the entire mindset, guidance and enforcement on issues under discourse in the western institutional domain. Yet the same unitary management system is not necessarily epistemologically sensitive to other cultural domains and social realities. The biggest conflict today is the divide between the understanding of neo-liberalism as the western belief and the Islamic Law among the Islamicists. The much needed bridging and dialogue between the divided worlds will continue as the most significant socio-scientific issue for all peoples for all times (Sardar, 1988).

The absence of the unique and universal epistemological premise in management neurocybernetic theory, one that can guide and be beneficial and acceptable to most of humanity remains the basis of global disorder. The urgency in closing up this gap with mutual understanding would be the proper direction for developing a management neurocybernetic systems model for the common wellbeing of all (Choudhury, 1996).

The quest for this universal and unique epistemological worldview must be both serious and reasoned. In the area of Computing Reality, the calculus of wellbeing requires the prevalence of discursive mechanism combined with the technocratic developments that can put analytical substance and application for such an approach in management neurocybernetic theory. Indeed, morality, ethics and values are not numinous entities of systems. They are equally as much measurable ones as are any material socio-scientific variable. See Einstein (undated), Choudhury (1995). Bohr (1985) mentions about his correspondence with Einstein on the scientific epistemological nature of ethics. In neurocybernetic theory of management systems with the universal epistemological model of unity of knowledge, the systemic endogeneity of morality, ethics and values (Choudhury, 1995) would be a distinct way of formalizing, measuring and implementing the interactively integrated and dynamics roles of these human imponderables in socio-scientific decision-making, human wellbeing and global future.

PROBLEMS OF INSTITUTIONALISM

The above ones are some examples at the social, political and corporation levels where transitivity of decisions and preferences is not found to work. The possibility of transitively expected inferences of decision-making failing to work can be generalized to all cases of complex decision-making where dynamic and interactive factors disturb a linear way of looking at the complex hierarchical decision-making process in organizational setting. What is true of organizational behavior is equally true of institutional situations. Institutions are broader organic forms and carry encompassing worldviews that can subsequently shape organizations (Feiwel, 1987).

30

For instance, the institutionalism of Public Choice Theory argues in favor of instilling the arguments and behavioral ramifications of methodological individualism into politics, constitutions and social contracts (Buchanan, 1971). Here the organizational theory is governed by the precepts of institutionalism being premised on neoclassical utilitarian theory (Hammond, 1989). On the other side, Global (International) Political Economy is a neo-Marxist perspective in critical examining the prognosis of capitalism in globalization process of conflict between power, politics, ownership and distribution (Palan, 2000). Here organizational behavior is governed by alternative forms of organizational awareness with an epistemological perspective contrary to global capitalism and global utilitarianism (Ruggie, 2002). New Institutionalism like Public Choice Theory weighs institutional influences on the behavior of individuals and households. Such a social transformation is further extended to the political and social fronts as well. But the causation that takes place between the 'institutional environment' and its human domain of influence pits a push and pull between neoclassical norms of profit-maximizing firms with constraints on transaction cost and the consequential transformation of social contracts and institutional behavior in the same light of neoclassical rationality. Here too we find the overarching power of institutionalism as a way of governing organizational behavior despite what transitivity of preferences may imply socially.

Combining the theory of institutionalism with organizational behavior in hierarchies of complex decision making projects the role of systemic and cybernetic thinking. This appears as feedback causality between human preferences, values, ethics, cultural vintages and the emanating embedded environments. By such causality the human minds are fashioned and the community and global order then operates in a reasoned, discursive and unified way.

Other writers quoted above sought to extend the cybernetic way of thinking to the domain of logical thinking, computer logic and several socio-scientific directions. We will examine these contributions now to establish the fact that the mindset of reasoned systemic discourse is a pervasive phenomenon. It must therefore be captured as a socio-scientific theme in formalizing a 'universal' model of the phenomenology of unity of knowledge in both machines (artifacts) and the human order.

CYBERNETIC LEARNING PARADIGMS

Autopoeisis Theory of Learning Systems

Johannessen (op cit) provides an incisive study on how a social system borrows from its environments to learn in two interconnected ways. Firstly, there is the positivistic loop that completes a certain task at the disposal of the social organization according to generated norms. Secondly, every such completion is simultaneously connected with the normative loop that cognizes an evolutionary learning system to continue on the process in repeated loops.

The author describes such a system in the following way: He uses the ideas of Luhmann (1986) and Maturana and Varela (1980) to explain Autopoeisis, as self-producing learning system in which two things get automatically regenerated by circular causation in order to maintain the self-production pattern. The pattern is defined firstly by its closed (normative) and open (cognitive) relations. These relations are structurally linked causing thereby an interaction to continuously appear between the two forms of relations that together define the self-producing autopoeitic learning. Hence the part on closure, that is the endowing of norms to the system, is necessary for processing the part on openness in cognizing the solutions in evolutionary learning paths. Such a relational interaction between the learning parts of the social system is viewed by Johannessen to comprise the phenomenology of all social system. Indeed the co-evolution of the two parts of autopoeitic learning in systems is the mark of circular causation relations. The author writes (op cit, edited, p. 359):

> In systemic thinking an important question is: What is the pattern which combines a given phenomenon or problem? The noticeable thing about patterns is that it is difficult to pinpoint cause and effect, i.e. there is a tacit dimension involved. A pattern can metaphorically be regarded as a circle or a spiral, and a circle has no beginning or end. The pattern is connected by relationships............

In later chapters we will develop at length and apply the methodology of circular causation in unified relations between continuously learning entities of embedded systems. In fact, we will show that circular causation is the methodological corollary of unified systemic relations, given the goal of attaining and recycling unity of knowledge between entities in such systems. Such entities are variables and agents characterizing the embedded and unified systems. Consequently, no system is truly independent of another. They form an interactive and integrated learning family of sub-systems. The capacity to learn continuously also characterizes such embedded systems and their entities and relations in inducing morality, values and ethics as the natural cause and effect of continued learning in unity of knowledge. The unity of knowledge is the endowed normative premise (closure) and the discursive learning process as opposed to imposing the norm is the positivistic content (openness).

The cybernetic nature of social system understood in the context of circular causation relations defining their embedded interconnections is now obvious. Cybernetic is a way of scientifically understanding the complexity of reasoned learning within and across embedded systems. The abiding consequence of such learning is that machines and social organizations as well as the field of institutionalism, which we studied above, are together emulating a unique learning pattern. That is systems are now self-producing and self-organizing wholes conveying normative rules combined with positivistic applications. In this definitional context, scientific reasoning, machine language (computer) and the human order are all seen as being endowed by the intrinsic property of circular causation and epistemic completeness.

Yet one point in the field of epistemological unity of knowledge that cautions the design of the learning patter between closure and openness is this. The phenomenology of cybernetics that is uniquely entrenched in all forms of learning systems would be impossible if diverse sub-systems as part of the overarching family have different norms. If this was the case, different norms or undefined normative field will yield different kinds of epistemology to the sub-systems in a family of systems. Consequently also sub-systemic and problem-specific positivism would not infer from any unique nature of the normative primal. Each sub-system would thereby degenerate into learning only intra-systemically. The inter-systemic learning and problem-specific purpose conveyed by the axiom of unity of systemic knowledge and its consequential problem-specific inferences to be derived, withers away. Consequently, on this point of selecting the primal norm the design of scientifically reasoned thinking on unity of systemic knowledge to solve particular problems and make positive inferences thereby, the norm assignment must be well determined. If this is not done, Johannessen's cybernetic argument concerning machine logic (computer) and social systems can be contested, despite his incisive desire to have holistic behavior between systems.

The missing specification of the fundamental norm to guide all unified systems, and thereby address their specific problems, is a perennial problem of scientific philosophy. It is markedly manifested in Darwin's evolutionism theory caused by mutations (Dawkins, 1976). Myrdal's (1968) idea of the 'wider field of valuation' leaves the determination of the unique norm to the philosophy of pluralism rather than *diversity* of representations of the unique norm, once it is specified. The problem of indeterminateness is also found in Popper's (1987) notion of the conjectural universe.

In our work, the critical norm will be assigned precisely according to the episteme of unity of knowledge that overarches systems by the search for 'universality' in learning systems. Thereby, both the normative and positive aspects of learning systems are interactively unified (integrated) along recursively generated co-evolutionary learning paths. Systems are nexus of such learning co-evolutions. Cybernetic theory now becomes a study of deciphering the learning behavior of such unified and co-evolved nexus of relational epistemology (Thayer-Bacon, 2003).

Methodological Differentiation in Paradigm Theory

Methodological complementarism in cybernetic literature (Yolles, op cit) is a concept used quite differently from our meaning of learning linkages by circular causation in unified systems of all kinds. The terms used in the literature mean differentiation and plurality in the cognitive perceptions of different worldviews in the systems perceivers. Systems paradigms as well as the approach to problem solving thus become differentiated. The role of belief, culture, governance, group interests and goals, social and economic contests and alienation and power structures all bundle themselves in groups to cause differentiation to take place. This is the mindset underlying the concept of methodological complementarism. Our usage of the principle of complementarities aforementioned is thereby opposite to the idea of complementarism in the cybernetic literature.

Methodological complementarism is exemplified by the physicist's understanding of the universe in ways contrary between the perspectives of General Relativity Theory and Quantum Theory. This divide in theoretical physics has been the sticking problem of Grand Unified Theories. Yet the quest for the theories of everything has persisted over a long time now (Hawking, 1985, Barrow, 1991). The underlying conceptual problem is reflected in the fact that the universe is intrinsically unified by the most reduced laws of everything. But the quest for this unique and universal law of unity of knowledge could not be identified and discovered in theoretical physics thus far.

In searching for the unified theory of systems and cybernetics Yolles (op cit) proposes a meta-systems methodology of complementarism. In his words (p.542): "Firstly, we propose that the problem of paradigmatic incommensurability within methodological complementarism can be resolved by not trying to consider the cognitive space of the paradigm, but rather to concentrate on the cognitive purposes of methodologies that must match the cognitive purposes of an inquiry."

Systemic Over-Determination in Marxist Political Economy

Likewise, in the theory of political economy we encounter the problem of over-determination of social epistemologies according to Marx (Resnick & Wolff, 1987). The authors explain the over-determination problem of systemic plurality based on intrinsic systemic conflicts and multiplicity rather than unification of knowledge. The problem in Marx's over-determination problem arises from the inability of the social, political and economic system with plethora of many relations to co-determine them. No binding norm exists in the over-determined system. The system-specific problem is then unsolvable and the over-determination problem leads into contradictions. The empty circularity of all internal relations distinctly from all other systems leads into reinforcing the conflicts between the differentiated systems while leaving the problem of solving any particular issue unresolved. Resnick and Wolff write (p.5):

> The contradictoriness of the process of theory appears on two levels. At one level there are differing and often conflicting theories, that is, sets of concepts. Practitioners within each of these spend part of their time in *criticism*, which we understand as chiefly the specification of differences (contradictions) between their own and other theories. At another level each distinct theory or set of concepts is itself contradictory in the sense that tensions and conflicts exist among the concepts comprising the set. Practitioners within any particular theory typically spend part of their time identifying and seeking to resolve those contradictions within their theory that they can recognize as such.

Another View of Methodological Complementarism and the Alternatives

Another concept of methodological complementarism used in cybernetic literature is similar to methodological individualism, which thoroughly defies the possibility for unity of knowledge between systems. Differentiated systems and individuated learning are promoted. At best a temporary interactive phase leads into mutations caused by competition, alienation and myopia. We have mentioned some such differentiated systems in the neoclassical economic category to be neo-liberalism as pronounced by Buchanan (op cit). In fact, all neoclassical economic systems are differentiated ones, due to their abiding assumptions of economic rationality, scarcity of resources, competition, transitivity of preferences and marginal substitution between alternatives. We have also pointed out the nature of embedded systems as opposed to differentiated systems in economy and society (Holton, 1992). But here too there is a substantive difference in understanding the embedded nature of interactive systems in organically unified systems caused by learning, from those conceptualized by Polanyi (1944). We will study later on that

37

Polanyi's embedded economic system is subsumed within the social system, giving the latter the governing power over the economy. Contrarily, learning embedded systems will be studied in our work as interactive, integrative and evolutionary process-based ones that learn mutually, continuously and across continuums (nexus) of entities and their complementary relations.

In this book we will dwell on the concept of the 'universal' in paradigm shifts to arrive at the concept of the worldview. The worldview is acceptable to all paradigms in the moral, ethical and social sense. This is the generalized conception of unity of knowledge in the sense of embedded learning systems together with their entities and relations. Indeed, unity of knowledge as the unique methodology of Grand Unified Theory in physics, of pansystemic unification in cybernetics, and embedded learning social systems is that universal norm, which can establish the epistemology of the 'universal' paradigm Such a universal paradigm governs all methodologies, while recognizing the diversity of systems, their problems, issues and the defining entities. Computing Reality imbibes in such a 'universal' theory of systems and cybernetics.

OTHER STUDIES IN EMBEDDED AND LEARNING SOCIO-SCIENTIFIC SYSTEMS

Socio-scientific system is a terminology we use here to mean pansystemic complementarities attained by means of circular causation arising from the central role of unity of knowledge. Because such a unique epistemology is intrinsic across most social, economic and cybernetic theories, and can be proven to be the 'universal' across unified system theory, therefore, we use the term socio-scientific for such epistemologically unified systems. In this sense therefore, a machine is as much a learning artifact by virtue of its two properties as is a human system. The two venues of scientific construction influenced by unity of knowledge as the common episteme are firstly found at the conceptual level. This characterizes the moral, ethical and value perspectives of constructing machine logic, as in cybernetic theory (also computer logic). We refer to such a moral, ethical and value theorizing at the conceptual level of reasoned scientific logic entering artifacts in the form of a socio-scientific theory of continuous machines (Choudhury, 1998). Secondly, socio-scientific reasoning premised on unity of knowledge as the episteme finds roots also in the social application of logical artifacts. While the second perspective is readily understood as the social positioning of artifacts, much less has been accomplished in the first area mentioned here. Cybernetic knowledge premised on unity of knowledge must devote to this area of scientific development in ethical machines (e.g. artificial intelligence of computer logic).

Historistic Socio-Scientific Systems

In the first of the socio-scientific constructions we examine two examples. The first one is concerning a theory of historicism, the understanding of causations in systems change as a particular example of historical entity. Kurt Hubner's (1983) thesis on historicism as a systems ensemble can be understood as an interesting study in the study of historistic phenomena from a cybernetic perspective. Another rather non-conventional understanding of cybernetic conception in the history of philosophical thought is given by the Islamic scholastic, Abdul Hameed [Imam] Al-Ghazali (trans. Buchman, 1998). We examine these from the non-conventional viewpoints of socio-scientific cybernetic potential in order to lay down the building blocks of the wider application of the theory of systems and cybernetics in the broader context of the human order interrelating with abstraction as an inner core of scientific thought.

Bertrand Russell once said that abstraction is the soul of practical power, and that it is strange we know little and yet we know so much. So little a knowledge, gives us so immense a power. Thus the field of abstraction is the soul of the paradigm shift in scientific reasoning. Also along a similar line of explanation Thomas Kuhn (1970) wrote about the structure of scientific revolution, that a revolutionary paradigm is a break away from the cumulative pattern of thinking in normal science. It brings about gigantic leap in human understanding of reality. This is often non-cumulative in nature and often does not establish a linear continuity with the prevalent thought. A good example in this regard is the theory of Quantum Physics that broke away from Relativistic and Newtonian thinking in theoretical physics. Likewise General Relativity was a break away from the linear conception of energy, forces and matter in Newtonian Physics. Both Relativity Physics and Quantum Physics are deeply philosophical in their argumentation. This is supported by the philosophical thought of Albert Einstein (Margenau, 1951) and Werner Heisenberg (1958). The relevance of deeply philosophical, if not metaphysical, groundwork of scientific reasoning at the epistemological and ontological levels is thus an established fact.

Indeed, the pursuit of scientific reasoning led by theoretical physics and cybernetics in the socio-biological study of symbiosis is toward discovering and defining the ultimate unity of the forces and relations of nature. Upon this premise could be constructed the explanation of all world-systems. On this point Hawking, (1988, p. 10) wrote:

> The eventual goal of science is to provide a single theory that describes the whole universe. However, the approach most scientists actually follow is to separate the problem into parts. First, there are the laws that tell us how the universe changes with time. ..... Second, there is the question of the initial state of the universe. Some people feel that science should be concerned with only the first part; they regard the question of the initial situation as a matter for metaphysics or religion.

Hawking acknowledges the role of the divine deity in the creation of the universe according to its laws. But he also treats such a divine action as a static phenomenon, which once caused has not been intervened with by divine will. Hence, in a cybernetic perspective of world-systems including cosmology, God ceases to play a dynamic role in creation. Consequently, the static and the once-and-for-all divine role in creation, negates the cybernetic nature of learning world-systems. This does not help us in generalizing the theory of cybernetic across learning complementarities between all systems and their entities including the dynamic role that the divine order plays in the governance of the universes with their entities having purpose and explanatory meanings.

To establish a generalized theory of cybernetics with its overarching nexus of learning systems and entities we turn to a combination of the theories presented by Hubner concerning systems ensemble and Imam Ghazali's conception of the learning universe under the guidance of the universe by divine laws. Such an integrated way of understanding learning systems spanning over complementing dimensions of knowledge, time and space would point toward the need and structure for a generalized theory of cybernetics having pervasive complementarities between the systems and their entities.

Kurt Hubner's Historistic System-Ensembles

Hubner's explanation of *historical science* in the form of system dynamics has three components. Firstly, he explains *historical system* as the collective co-evolution of given scientific thought over time. Though, along such an evolution of scientific thought the old ideas may not always be repeated. Even with the retention of the same terms, the old terms could change in their meanings, formulation, and thereby application along the evolution of historical system. An example here is the understanding of the meaning of space. It has acquired substantively different meanings in scientific understanding over historical evolution of systems of thoughts. Newtonian motion-like dynamics is simple in form, having to do with the motion of particles in void and non-void spaces. Einstein's large-scale relativistic spacetime structure, in which space is the result of the simultaneous occurrence of light and event, defines the concept of the non-viability of a field (space) in physics in the absence of matter (Einstein, 1954). Quantum Physics is the study of sub-atomic particles and their occupied spaces. This is the micro-physical structure of events controlled not by the incidence of light on particles. Rather Quantum space is defined by the pervasive existence of probabilistic interaction between position and momentum of moving particles (Kafatos & Nadeau, 1990).

From the concept of historical systems in perpetual evolution to higher levels of complexity caused by rules and axioms governing various ideas, Hubner next draws his concept of *historical system-ensembles*. System-ensembles are formed by the collectivity of rules governing all forms of ideas. System-ensembles reflect momentary spaces of collected theories, concepts and ideas that define the development of given scientific fields. System-ensembles are thereby cross-sections at given scientific epochs along the movement of historical systems. In Hubner's words (op cit, p. 109): "Scientific systems, that is, theories and hierarchies of theories, as well as the rules governing scientific work, are thus all a part of the collective ensemble that presents us with the world of rules within which we live and act at any given time."

The combination of Hubner's concept of historical systems and that of historical system-ensembles yields his concept of *historical situation*. Hubner (op cit, p. 109) defines it as "a historical time period (*Zeitraum*) that is dominated by a particular system-ensemble."

Any system-ensemble remains permanently unstable due to the absence of

42

some given unifying relation between all the rules governing different ideas and theories inside a historical system within an ensemble. This can be seen as the dialectic of change and evolution of historical systems, a notion that is of Darwinian and Popperian category (Popper, op cit). Consequently, the evolution of system-ensembles over historical systems in time is marked by growing effort to eliminate or reduce the degree of instability, incoherence and inconsistencies between theories and ideas within system-ensembles included in the collectivity of historical systems at any moment of historical evolution of all theories and ideas.

The challenge to Hubner's systems perspective of historical science is to discover such a rule that while it supercedes and interconnects all contending theories and ideas removing the inconsistencies, it also establishes stability of the system-ensembles. Yet Hubner denies this property of learning to his concept of system-ensembles. In his words (op cit, p. 110), "Now, since the ensemble has a structure in virtue of the relations obtaining between its elements, it might be possible to suppose that perhaps all these elements are deducible from one fundamental element of the ensemble. But this is simply not the case." Hubner's concepts of system-ensembles and historical situation are not driven by any uniquely universalizing core relations that regulate all relations despite their diversity and problem-specificity in various disciplines and contexts.

This missing link of a unique core episteme that can unify all relations in Hubner's historical situation can be completed by the way that Imam Ghazali viewed the historistic process, and within it, all world-systems. Imam Ghazali (1998, p. 16)"

Existence can be classified into the existence that a thing possesses in itself and that which it possesses from another. When a thing has existence from another, its existence is borrowed and has no support in itself. When the thing is viewed in itself and with respect to itself, it is pure nonexistence. It only exists inasmuch as it is ascribed to another. This is not a true existence, just as you came to know in the example of the borrowing of clothing and wealth. Hence the Real Existence is God, just as the Real Light is He.

Imam Ghazali's Theory of Existence and Reality in Learning Systems

The pervasive nature of the unity concept in all of history seen as evolutionary systems, including the human and non-human order upon which Imam Ghazali constructs his understanding of existence, is mentioned in the Qur'an (41:11, 12):

> Moreover He comprehended in His design the sky, and it had been (as) smoke: He said to it and to the earth: "Come you together", willingly or unwillingly." They said: "We do come (together), in willing obedience."

> So He completed them as seven firmaments in two Days, and He assigned to each heaven its duty and command.......

The above Qur'anic verses bring out the context of systems and the need for order, objective and sustainability in them. Imam Ghazali refers to such systemic features in the context of his explanation of the meaning of existence. The meaning of existence according to the concept of historistic system is possible only in the context of unity of the divine laws. In the above Qur'anic verse, giving the example of the early cosmos, the historical situation in unity of knowledge is extended over time, spatial continuums (nexus), dimensions, varieties of systems, and their entities. All are unified together by the divine laws to realize their meanings and existence. Without this epistemic core of unity of divine knowledge no system can continue, acquire sustainability, or be defined in the sense of social and reasoned order with objective criterion goals.

Contrasting Hubner's unstable and inconsistent, open-ended and incom-

44

plete idea of historistic system-ensembles with Ghazali's holistic concept of learning systems invokes the search for a generalized and 'universal' theory of systems and cybernetics. Such a theory should be able to answer all systemic forms, though with different and contrasting meanings. That is, if the differentiated system theory, even in its limited form of Hubner's system-ensembles, is false, then the generalized and 'universal' theory of systems and cybernetics must be able to analytically explain both the false systems and their opposite, the unified worldview. The same theory must be capable of explaining historistic system theory from all perspectives across time, space and dimensional continuums as nexus of knowledge and knowledge-induced entities and their relations.

TOWARDS A GENERALIZED THEORY OF SYSTEMS AND CYBERNETICS IN THE LIGHT OF UNITY OF KNOWLEDGE

A pansystem theory was recently tried by Xuemoue & Dinghe (op cit, p. 679). They define a generalized theory in the pansystems mould as "a combinative investigation, extension, a complement of cybernetics within the framework of pansystems theory. Certain pansystems regeneralization or generalized optimization of the categories, principles and theorems of cybernetics and generalized control, observation, communication, systems (fuzzy systems), information and entropy, etc. are realized, including certain new explorations to philosophy, mathematics and relativity." It is interesting to note the concept of generalized observations in pansystems theory. Generalized observations are inter-entities and inter-systemic processes that make these interact. Yet the set of such observations with respect to specified problems under investigation must be bounded by control. This is the concept of observocontrollability and is specific in relation to given issues and problems. Thus, the concept applies only to specificity of the issues and problems under study. The numinous idea of universal connectivity between *all* things that which Capra (1983) refers to is discarded because of the impossibility to study any such extensively relational problem by means of limited knowledge-flows in any space-time situation within the unification experience that cannot *completely* realize all unified system-ensembles. No optimization of the learning systems in any shape and form is therefore implied in the epistemology of unity of knowledge in learning systems.

We can now think of a pansystemic generalized system and cybernetic theory as a system of pervasive complementarities existing between learning systems, their entities and the circular causation of entity-relations (ontology) within the limits of observocontrollability. Now Hubner's specificity of ideas and theories in system-ensembles applies for addressing given problems and issues. But the internal instability and inconsistencies of any such system-ensemble is removed by introducing the episteme of unity of knowledge as circular causation for unifying the entity variables by means of the epistemic axiom.

Hence comes in the relevance of Imam Ghazali's formulation of learning systems over knowledge-induced space, time and continuum of dimensions. But the issues and problems remain specific in nature, not numinous. The above-mentioned Qur'anic verses refer to the specific theme of cosmology but extend the underlying episteme of unity of the divine laws to all particulars of creation. This is the idea of Reality and Existence within the limits of observocontrollability as explained by Imam Ghazali.

The historistic system-ensembles now become unified world-systems by virtue of learning on the basis of the episteme (axiom) of unity of knowledge prevailing for each and all systems and their entities and relations. Consequently, the gaps caused by conflicts, inconsistencies and instabilities between theories and ideas within system-ensembles are replaced by restructuring them with learning on the basis of selection and design of the systems by means of unity of knowledge. The axioms, objectives, methodology, driving instruments and inferences premised on unity of knowledge now become central in the systemic reconstruction. The generalized and universal theory of systems and cybernetics is now established by the worldview of unity of knowledge in evolutionary historistic fields of time, space and continuums with specificity of the problems and issues under investigation. We will examine such a generalized and universal theory of systems and cybernetics as we proceed on in this book. Computing Reality as the manifestation and career of this worldview of unity of knowledge in systems will be projected in specific issues and problems of the socio-scientific order and computer logic.

SUSTAINABILITY AND COMPUTING REALITY

The generalized and universal theory of systems and cybernetics for socio-scientific problems must be characterized by pervasive complementarities between systems and their entities and entity-relations according to unity of knowledge as the epistemic axiom. This is methodologically accomplished by circular causation between the learning and unifying observocontrollable variables, their objective criteria and the underlying axioms. The entire socio-scientific investigation is thus based on three inherent parts that together comprise the scientific reasoning of Computing Reality.

Firstly, the generalized and universal theory of systems and cybernetics in Computing Reality for every particular issue and problem is axiomatically governed by the epistemology of unity of knowledge. Secondly, the episteme leads to the formal methodological conceptualization of the system within which specific issues and problems are to be investigated. This constitutes the ontology of the theory construction in the light of the epistemology of unity of knowledge.

Thirdly, the epistemology and ontology combine to yield to empiricism, institutionalism and inferences of the analytical investigation at hand.

The generalized and universal theory of systems and cybernetics, and now the derived meaning of computing reality in its fold, lead to the concept of systemic sustainability. In this chapter these concepts have been developed in the light of socio-scientific systems by a review of the literature. In Chapter 3 the same conclusions will be found to hold for an application of computer model of computing reality. Hence the concept of systemic sustainability remains overarching between socio-scientific and computer systems within the fold of epistemology of unity of knowledge.

Systemic sustainability means historical evolution of learning systems and their entities and entity-relations that move across system-ensembles and that are unified intra- and inter-systems by the law of unity of knowledge. The comprehensive delineation of such learning systems as sustainable ones comprises the methodology emanating from the sequential totality of the epistemology of unity of knowledge, its ontology in formalism and these together enabling quantitative, empirical and inferential results to be derived and applied. The persistence of sustainable systems evolves over space, time and dimensions of knowledge, all premised on the unification experience and a holistic purpose. Within this holistic domain are included the human and non-human, the real and abstract entities as particulars for investigation by means of the theory of generalized and universal systems and cybernetics as the methodology underlying Computing Reality.

CONCLUSION

Continuing on from Chapter 1, the theme of Computing Reality is seen in chapter to be embedded in the conceptualization, formalism and measurement followed by inferences concerning the neurocybernetic theory of learning systems. Learning is embedded only in those systems that emanate from and simulate through all issues under discourse based on the premise of unity of knowledge. The review of the literature in this chapter showed contrarily, that theories of systems and cybernetic are not necessarily entrenched in the epistemology of unity of knowledge. Thus, even the central principle of pervasive complementarities between systems, their entities, agents and relations is contrarily perceived in the theory of methodological complementarism and unity of knowledge as explained by interactively developed interactive and evolutionary world-systems.

In this chapter, we have noted that much of the theory of systems and cybernetics as conceptualized within the socio-scientific domain with multidimensional imponderables, must be re-thought along lines of the episteme of unity of knowledge. Without this fundamental revolution in the mindset and application of the new neurocybernetic worldview, the theme of computing reality would not be able to incorporate the moral and ethical values in it. The building blocks of a comprehensive, unique and universal worldview underlying a theory of systems and cybernetics with the goal of attaining social wellbeing will remain distanced. Thus our objective as pointed out in Chapter 1 is to search for, derive, formalize and apply the emergent theory of systems and cybernetics premised in the fold of unity of knowledge to all learning systems.

REFERENCES

Al-Ghazali, A.H. [Imam] trans. Buchman, D. 1998. *The Niche of Lights*, Provo, Utah: Brigham Young University Press.

Ansari, J. 1986. "The nature of international economic organizations", in *Political Economy of International Economic Organizations*, Boulder, CO: Rienner, pp. 3-32.

Arrow, K.J. 1951. *Social Choice and Individual Values*, New York, NY: John Wiley & Sons.

Barrow, J.D. 1991. *Theories of Everything, the Quest for Ultimate Explanation*, Oxford, Eng: Oxford University Press.

Bohr, N. 1985. "Discussions with Einstein on epistemological issues", in H. Folse, *The Philosophy of Niels Bohr: The Framework of Complementarity*, Amsterdam, The Netherlands: North Holland Physics Publishing.

Bolton, D. & Hill, J. 2003. *Mind, Meaning, and Mental Disorder*, Oxford, Eng: Oxford University Press.

Buchanan, J.M. 1971. *The Bases for Collective Action*, New York, NY: General Learning Press.

Capra, F. 1983. "The systems view of life", in his *The Turning Point*, London, Eng: Flamingo.

Choudhury, M.A. 1995. "A mathematical formalization of the principle of ethical endogeneity", *Kybernetes: International Journal of Systems and Cybernetics*, 24:5, pp. 11-30.

Choudhury, M.A. 1996. "Economic integration in the Sextet region and the Middle East Peace Accord", *International Journal of World Peace*, 13:2, pp. 67-78.

Choudhury, M.A. 1998. "A socio-scientific theory of continuous machines", *Cybernetica*, XLI: 2-4, pp. 251-271.

Choudhury, M.A. 1999a. "A philosophico-mathematical theorem on unity of knowledge", *Kybernetes: International Journal of Systems and Cybernetics*, 28:6/7, pp. 763-776.

Choudhury, M.A. 1999b. "Global megatrends and the community", *World Futures*, Vol. 53, pp. 229-252.

Condorcet, M.J. & Caritat, A.N. 1785. *Essai sur l'application de l'analysis a la probabilite des decisions rendues a la pluralite des voix*, Paris, France : Impremerie Royale.

Cummings, J.F. 2005. *How to Rule the World, Lessons in Conquest for the Modern Prince*, Tokyo, Japan: Blue Ocean Press, Aoishima Research Institute.

Dawkins, R. 1976. *The Selfish Gene*, Oxford, Eng: Oxford University Press.

Debreu, G. 1963. "A limit theorem on the core of an economy", *International Economic Review*, 4:3, pp. 235-246.

Einstein, A. 1954. "The problem of space, ether, and the field in physics", in Commins, S. & Linscott, R.N. eds. *Man and the Universe: The Philosophers of Science*, New York, NY: The Pocket Books, Inc.

Einstein, A. undated. "The laws of science and the laws of ethics", in his *Essays in Physics*, New York, NY: Philosophical Library.

Emmerij, L. 2000. "World economic changes at the threshold of the Twenty-First Century", in Pieterse, J.N. ed. *Global Futures: Shaping Globalization*, London, Eng: Zed Books, pp. 53-62.

Feiwel, G. Ed. 1987. *Arrow and the Foundations of the Theory of Economic Policy*, London, Eng: Macmillan.

Hammond, P.J. 1989. "On reconciling Arrow's theory of social choice with Harsanyi's Fundamental Utilitarianism", in G.R. Feiwel Ed. *Arrow and the Foundation of the Theory of Economic Policy*, pp. 179-221, London, Eng: Macmillan.

Hawking, S.W. 1985. "Is the end in sight for theoretical physics? Cambridge University Lecture, Reprinted in Boslough, J. *Stephen Hawking's Universe*, New York, NY: Avon Books, pp. 119-139.

Hawking, S.W. 1988. *A Brief History of Time, From the Big Bang to Black Holes*, Toronto, Ont: Bantam Books.

Heisenberg, W. ed. R.N. Anshen, 1958. *Physics and Philosophy*, New York, NY: Harper & Brothers.

Holton, R.J. 1992. *Economy and Society*, London, Eng: Routledge.

Hossain, M.S. Choudhury, M.A. & Mohiuddin, M. 1998. "Cybernetic interrelationships: an empirical study with respect to Bangladesh ecology", *Kybernetes: International Journal of Systems and Cybernetics*, 27: $&5, 485-495.

51

Hubner, K. trans. Dixon, P.R. Jr. & Dixon, H.M. 1983. *Critique of Scientific Reason*, Chicago, ILL: University of Chicago Press.

Inglott, P.S. 1990. "The rights of future generations: some socio-philosophical considerations", in S. Busuttil, E. Agius, P.S. Inglott & T. Macelli Eds. *Our Responsibilities Towards Future Generations*, pp. 17-27, Malta: Foundation for International Studies & UNESCO.

International Monetary Fund, 1995. *Our Global Neighbourhood*, New York, NY: Oxford University Press.

Jackson, M.C. 1993. Systems *Methodology for the Management Systems*, New York, NY: Plenum Press.

Johannessen, J.A. 1998. Organization as social systems: the search for a systemic theory of organizational innovation processes", *Kybernetes: International Journal of Systems and Cybernetics*, 27:4&5, pp. 359-387.

Kafatos, M. & Nadeau, R. 1990. "The road untraveled: enlarging the new logical framework of complementarity", in their *Conscious Universe*, New York, NY: Sringer-Verlag.

Kuhn, T.S. 1970. *The Structure of Scientific Revolution*, Chicago, IL: University of Chicago Press.

Luhmann, N. 1986. "The autopoiesis of social systems", in Geyer, F. & Van der Zouwen, J. eds. *Sociocybernetic Paradoxes*, Beverly Hills, CA: Sage, pp. 172-192.

Machiavelli, N. trans. Dionno, D. 1966. *The Prince*, New York, NY: Bantam Books.

Margenau, H. 1951. :Einstein's concept of reality", in Shilpp, P.A. ed. *Albert Einstein, Philosopher-Scientist*, New York, NY: Tudor Publishing Co.

Maturana, H.R. & Varela, F.J. 1980. *Autopoiesis and Cognition: The Realization of the Living*, Reider, Dordrecht.

Minogue, K. 1963. *The Liberal Mind*, Indianapolis, IN: Liberty Fund.

Mommsen, W.J. 1989. "Max Weber on bureaucracy and bureaucratization: threat to liberty and instrument of creative action", in *The Political and Social Theory of Max Weber*, Chicago, ILL: The University of Chicago Press, pp. 109-120.

Morgan, M.L. & Quarter, J. 1991. "A start-up experience: the case of the Big Carrot", in `Wisman, J. D. ed. *Worker Empowerment, the Struggle for Workplace Democracy*, New York: The Bookstrap Press, pp. 145-158.

Myrdal, G. 1968. "The wider field of valuations", in *Asian Drama, an Inquiry into the Poverty of Nations, Vol. 1*, pp. 49-127, New York: Pentheon.

Palan, R. 2000. "New trends in global political economy", in Palan, R. ed. *Global Political Economy, Contemporary Theories*, London, Eng: Routledge, pp. 1-18.

Parsons, T. and Smelser, N. 1956. *Economy and Society*, London, Eng: Routledge and Kegan Paul.

Polanyi, K. 1944. *The Great Transformation*, New York, NY: Rinehart.

Popper, K. 1987. "Natural selection and the emergence of mind" in G. Radnitzky & W.W. Bartley, III Eds. *Evolutionary Epistemology, Rationality, and the Sociology of Knowledge*, pp. 139-155, La Salle, IL: Open Court, 1987.

Raymond, N. 1991. "The 'Lost Decade' of development: the role of debt, trade and structural adjustment", in Jackson, R.M. ed. *Global Issues 93/94*, Guilford, CON: The Dushkin Publishing Group, Inc. pp. 112-123.

Resnick, S.A. & Wolff, R.D. 1987. *Knowledge and Class, A Marxian Critique of Political Economy*, Chicago, ILL: The University of Chicago Press.

Ruggie, J.G. 2002. "Epistemology, ontology, and the study of international regimes", in his *Constructing the World Polity*, London, Eng: Routledge, pp. 85-101.

Sardar, Z. 1988. *Islamic Futures, the Shape of Things to Come*, Kuala Lumpur, Malaysia: Pelanduk Publications.

Sen, A. 1977. "Rational fools: a critique of the behavioural foundations of economic theory", *Philosophy and Public Affairs*, Vol. 6.

Simon, H.A, 1957 [1952-53]. "A comparison of organization theories", reprinted in *Models of Man*, New York: John Wiley & Sons, Inc.

Simon, H.A. 1957. *Models of Man*, New York, NY: John Wiley & Sons, Inc.

Simon, H. A. 1960 reprinted 1987. "Decision making and organizational design", in Pugh, D.S. ed. *Organizational Theory*, Hammondsworth, Middlesex: Penguin Books, pp. 202-222.

Singer, H. & Ansari, J.A. 1988. "The international financial system and the developing countries", in *Rich and Poor Countries*, London, Eng: Unwin Hyman, pp. 269-285.

Sklair, L. 2002. "Transnational corporations and capitalist globalization", in *Globalization, Capitalism and Its Alternatives*, Oxford, Eng: Oxford University Press, pp. 59-83.

Thayer-Bacon, B.J. 2003. *Relational "(e)pistemologies"*, New York, NY: Peter Lang.

Xuemoue, W. & Dinghe, G. 1999. "Pansystemic cybernetics: framework, methodology and development", *Kybernetes: International Journal of Systems and Cybernetics*, 28:6/7, pp. 679-694.

Yolles, M. 1998. "A cybernetic exploration of methodological complementarism", *Kybernetes: International Journal of Systems and Cybernetics*, 27:4&5, pp. 527-542.

CHAPTER 3: A COMPUTERIZED MODEL OF COMPUTING REALITY - A CASE OF ASSESSING SYSTEM RISK CAUSED BY FLOODING

Computing Reality is a rigorous study in the theory and application of the phenomenology of socio-scientific systems. The challenge in such a project is to compute aspects of the real-world facts by identifying interrelationships among objects, processes and events. Yet such a computation of reality requires the understanding of the logic of the problems under study. This leads to the correct formulation and understanding of the underlying concepts and epistemology pertaining to the problems and issues at hand. Through the phenomenological approach premised on a unique praxis of unity of knowledge, methodological differences between social and scientific problems are avoided in terms of their common epistemological premise and formalism. Such a formalism comprises the ontology, understood in the engineering sense, barring metaphysical confusions. Yet the issues under investigation in different systems remain diverse. Thereby, machine logic shares the same reasoning as socioeconomic issues, but address different and diverse issues and problems.

In terms of machine logic we define first the expansion stage. It consists of understanding the relationships among real world elements (objects, processes and events). Then there is the invention stage. It is concerned with the formulation of the relationship among the elements. This stage is followed by the realization stage. It consists of the development of machine logic to perceive socio-scientific formulation in a machine language format. In this way, the phenomenology of machine logic is constructed to address various socio-scientific issues and their methodological analysis.

However, the complete understanding of the construed phenomenology of machine language depends on the development of innovative computing logic, premised on a robust methodology. Such a methodology is encapsulated in learning processes comprising systemic interaction, integration and evolution (IIE) among the real-world entities, as found necessary to understand change and configuration in any dynamic state of the real-world problem investigation. The procedures for developing innovative computing logic explained here along with their applications to real-world problems will be demonstrated in this chapter.

A MODEL OF COMPUTING REALITY (MCR)

Modeling involves abstraction and conceptualization of a problem. These states are expressed either in quantitative or qualitative form (Turban and Aronson, 2001). They involve identification of the variables associated with the problem that is addressed. This is followed by the establishment of interrelationships between the variables appearing to explain and compute measurable indicators and impute logical meaning to other ones.

In this context, a computational model involves visualization of a real-world system in a 'computer-world'. The binary knowledge of the 'computer-world' logic is expressed in terms of binary digits ('0' and '1') through a continuous process of relational transformations. Such a transformation happens at three levels, namely, the conceptual, logical and physical levels.

At the conceptual level, the entities within a system and their relationships are identified in the form of an abstraction of reality. This level answers the question, 'what will the system achieve?' The answer at the abstraction stage involves an expanding experience, whereby more knowledge relating to the system problem domain is gathered (Layzell and Loucopoulos, 1989).

The knowledge at the logical level tries to answer the questions, 'how performance is achieved?' At this stage, formalism of a system model is designed. Examples of such models are linear programming model, relational data model, and raster and vector models. The latter ones are used to visualize spatial relationships of the entities existing on the surface of the earth. This logical level is sometime called the invention stage of the computational system model. However, realization of this stage depends on gathering enhanced information supporting the conceptual level of the computational model.

The task of the physical stage is to construct the logical level, which visualizes the model system for a 'computer world'. This stage represents realization of the model system (Hossain, 2005). Various computer language paradigms are considered at this stage. They enable the transformation of the logical model into binary machine code i.e '0' and '1'. An example of the physical level model is referred to as the 'shapefile' in spatial domain analysis, enabling the analyses of spatial cause-effect relations among the entities (Choudhury and Hossain, 2006). 'Shapefile' is a vector-based data format, containing the geometric location and attribute information of the objects.

Interlinking the three stages, namely the conceptual, logical and physical levels by circular causation between them generates a formal model of systems and cybernetics. Such a model becomes the essential groundwork of a Model of Computing Reality (MCR) in the perspective of the computer world. Such an MCR can allow the construction of a robust and reliable phenomenology of socio-scientific study. Such a formal problem is then investigated by means of machine language, i.e. in the computer-world perspective. As a special case, such a system model, in terms of computer-world language, can be explained by means of the problem of flooding in the phenomenological sense of computing reality. This idea is explained in this chapter.

However, the success of such a Model of Computing Reality (MCR), that is to visualize a real-world system in reference to the 'computer world', depends on the selection of an appropriate methodology. The methodology would yield choices of entities leading to the formulation of their interrelationships. These together set up the conceptual view of the MCR. Next they are followed by the development of the logical and physical views. Such a comprehensive methodology should work as an underlying guideline/philosophy to develop the above-mentioned three views considered in developing the MCR.

Such a methodology, at the conceptual level, should incorporate the knowledge-level of a system. This entails the epistemology of the MCR. This stage of the comprehensive methodology at the logical level should deal with what exists in the world (i.e. theory of being = ontology, understood here in its engineering meaning). Thereby, the ontology of the MCR is well defined.

The methodological stage at the physical level of the MCR should be concerned with the construction of phenomenology of machine logic and its application in solving real-world problems, an example of which that we will treat here is flooding. The application stage of the machine language, which is the ontological principle in MCR, represents the ontic (application, evidential) stage of the MCR. Yet, simply incorporating the above-mentioned three views (conceptual, logical and physical) in developing the MCR as an independent system, is not enough. A system and cybernetic delineation of the model requires interaction (I), integration (I), followed by their co-evolution (E) by way of learning between complementing entities through circular causation between such variables and their relations as formulated for the problem under study. That is, the methodology becomes one in relational epistemology.

RELATIONAL EPISTEMOLOGY

Relational epistemology has stages of phenomenological meanings. First, there is the axiom or a premise. We denote this original axiomatic premise by the absolute stock of knowledge, denoted by Ω. The stock of complete knowledge must be transmitted onto an ontological level of system explanation. Through such a transmission mechanism, the axiomatic premise is explained in real-world context. The result of the combination of Ω, while being mapped onto a real-world problem under study by a mapping say Φ, now excites the real-world investigation within a socio-scientific discourse medium, denoted here by Φ^*. In this way, we have the three components, (Ω, Φ, Φ^*) that go together in the formulation of the ontological stage of computing reality in the MCR. The MCR is now made ready to yield a conceptual view of the problem under study by the MCR.

The methodology of (Ω, Φ, Φ^*) enables the derivation of continuous knowledge-flows denoted by $\{\theta\}$ that arise endogenously through the interactive, integrative and evolutionary processes (IIE-processes) related to the circular causation between the selected variables denoted by $\{X\}$ in the model expressions that ensue. $\{X\}$ comprise both state and policy or instrumental variables. Since knowledge-flows $\{\theta\}$ induce the $\{X\}$-variables, we write these together in the functional form as $\{X(\theta)\}$-variables. Thus a bundle of knowledge-induced state and policy variables denoted by $\{X(\theta)\}$ emerge to design a solution to the problem through an IIE-process.

The evaluation of the IIE-processes in terms of attaining unity of systemic knowledge between the $\{\theta, X(\theta)\}$-variables is done by an objective criterion function. Since all problems carry a moral and ethical content, their objective criterion functions when compounded together may be called the Wellbeing Function, denoted by W(θ,X(θ)). The simulation of W(.) in terms of the circular causation between {θ, X(θ)}-variables denotes the physical/ontic view of the problem-solving in the MCR.

In the light of the IIE-process model of the MCR, Figure 3.1 illustrates the circular causation view. In this figure, the ontological principle represents a new universe of knowledge (i.e. the invention stage of the MCR) and the ontic stage represents the application aspect (i.e. the realization level of the MCR).

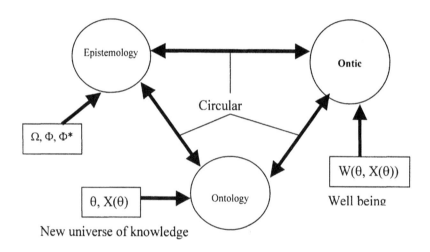

Figure 3.1: Circular Causation in an MCR Representation

Figure 3.1 gives an encapsulated view of the concept of circular causation, showing in it how cause-effect interrelationships among the three views (conceptual, logical and physical) can be illustrated as the principal components of a Model of Computing Reality (MCR). Such cause-effect interrelationships among the three views remain absent in a traditional computational model.

For example, the Turing way of computing reality consists of the identification of the processes such as 'turn left', 'scan', and 'turn right'. These are necessary to fill a square box of possibilities. They allows for an understanding of the change of state of the square box as the consequence of the mentioned processes. Although this technique shows the evaluation of new states resulting from the processes in the object box, the approach is unable to show the concept of circulation causation between the states in the object box. In addition, the knowledge-level view, which is necessary for understanding the interrelationship among the objects, remains absent. Hence, the logical view that defines computing reality is difficult to achieve. That is because the integration, interaction and evolution (IIE) among the three views (conceptual. Logical and physical) cannot be achieved. Consequently, the constructed phenomenology of the MCR system cannot be applied in deriving solution to real world situations.

In this chapter, the development and application of an MCR (based on relational epistemology) to assess the risks of flooding is presented. A description of the flooding problem is delineated. A discussion of the existing approach to assess risk due to flooding follows. A risk assessment approach is formulated in the light of the MCR concerning flood control. This problem domain in turn is used to develop conceptual, logical and physical views of the MCR.

FLOODING PROBLEM:
AN EXAMPLE OF PHENOMENOLOGICAL STUDY

Floods are one of the severest forms of natural hazards, causing extensive natural disasters (CRED, 1993). On a global scale, floods adversely affect 76% of the people of all those who are affected by natural disasters. Other natural hazards, such as hurricanes, typhoons and tornadoes combined affect only 18% of the people. The study of the phenomenon of flooding involves complex, interrelated and multidimensional geophysical processes. They interconnect causes of loss of lives with damage to property caused by inundation and further causing inability of communities to cope with floods. The adverse impact of flooding is realized when it disrupts road communication networks of a country. Such infrastructure networks provide the socioeconomic life-line to communities. To engage these interdependent causations, an MCR is developed in this chapter and applied to the problem of controlling flood-risk on route-networks.

Flood-Risk Assessment

The risk of flooding on national infrastructure depends on the identification of the probability of interaction between various flood dimensions (depth, area) and the infrastructure components (Hossain, 2000; Davies and Hossain (2001). The degree of this interaction can be measured both in qualitative and quantitative

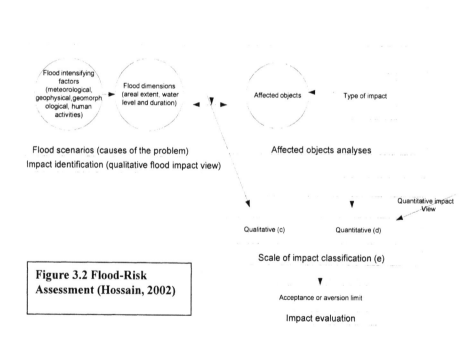

Figure 3.2 Flood-Risk Assessment (Hossain, 2002)

62

terms. The measurements here form a basis to identify the scale of risk classification on an object (for example, route network). Flood-risk is assessed.

The risk of flooding can be evaluated in the light of both the qualitative and quantitative classification of flood-risk. These assessments could provide insight to effective decision-making in reducing the flood-risk on a vulnerable object. Hence, there exist three stages to assess the risk of flooding, namely, identification, determination and evaluation. The above viewpoint of the risk assessment approach is encapsulated in Figure 3.2.

A computational model to assess the risk of flooding should take the three stages into account, especially in developing the conceptual model. That is because this will help to identify the entities related to flood intensifying factors, flood dimensions and flood affected real world objects (as shown in Figure 3.2), together with their interrelationships required for conceptualizing MCR.

However, the approach of flood-risk assessment as depicted in Figure 3.2 does not illustrate how risk-diversification is carried out. The reduction of risks by activating different policy variables is absent in this approach. Although it illustrates the interaction between flood and the objects, it is not able to show the interaction of flood with various policy variables, like installation of culverts on a route network, and increasing the water discharging capacity of river network or road level modification. The latter interaction will eventually form a circular causation. This means that the impact illustration of one variable on the other, between flood and policy variables, and thereby developing complementarities or integration through interaction between such variables will be possible.

Identification of appropriate policy variables will be carried out in a discursive manner between scientists, engineers, policy-makers and the community. Such a discursive process signifies the dynamics of relational epistemology. The immanent learning process is thereby induced by knowledge-flows. This is reflected in the policy variables. It can then be seen that systemic relational epistemology ensures identification of the variables/entities and their IIE-process interrelationships required to develop the appropriate conceptual view of the MCR. This aspect of system modeling is absent in the traditional computing model.

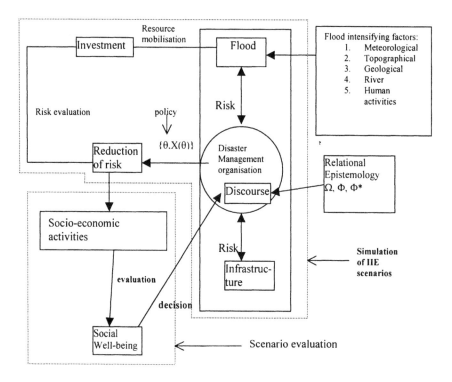

Figure 3.3: Problem Domain of the MCR to Assess Flood-Risk

The evaluation (E) of risk reduction is examined in terms of increasing socioeconomic activities in terms of the impact of the knowledge-induced policy variables included in the vector $\{X(\theta)\}$. By taking the above viewpoint into account a new approach to assess the risk of flooding is presented in Figure 3.3. This formulation presents the problem domain of the MCR.

64

Identification of the problem domain of such a model is the most important one, since the domain works as a starting point to start the expansion (epistemology/conceptual), invention (ontology/logical) and realization (ontic/physical) works in developing an MCR. However, it can be seen from Figure 3.3 that the output of the evaluation (E) component of the problem domain of the MCR must reflect the complementary linkages between the knowledge-induced variables in the light of relational epistemology. Such linkages result in new knowledge-flows following circular causation between the variables in terms of their formal computable relations. Experience generates new knowledge that in turn reveals new facts, rules, case studies and possibilities.

In the following section, a discussion on the framework of developing various models related to the three views of the Model of Computing Reality is presented.

THE FRAMEWORK OF SYSTEM DEVELOPMENT
LIFE CYCLE IN MCR

The System Development Life Cycle (SDLC) can be considered in a framework of developing various models at three levels (conceptual, logical and physical) of the MCR (Jeffrey et al., 1999; Kendall, 1996; Davies and Layzell, 1993; Bruch, 1992; Kendall and Kendall, 1988; Turner et al, 1987). The three-phase circular causation cycle provides an effective way of engineering such a model. In practice, there exist a number of stages in the development life cycle. They can be categorized under three different views including the conceptual, logical and physical phases.

The tasks of each view have to be carried out by using the problem domain (Figure 3.3) of the MCR to assess the impact of flooding. For example, computational concepts, such as the Standard Entity-Relationship (ER) modeling and data flow diagrams (DFDs), can be used to develop the conceptual view of the MCR. Since flood and infrastructure are spatial objects, and they are spatially interrelated, GIS (Geographical Information Systems) technology should be considered in constructing the ontic stage of the complete phenomenology of the MCR of flooding. The logical or ontological view of this MCR needs to be developed taking the schema of its constructive environment, which is the GIS. In the following section, the development procedures of the three views of the MCR will be demonstrated.

Conceptual View of the MCR

The components of the phenomenological perspective consist of entities, processes and events. An entity represents the real-world aspect of a model system. A process and an event represent the dynamic aspect of that system. The traditional entity modeling approach is widely used for modeling static aspects of a system (Flynn, 1998; Flynn Fragoso, 1996; Chen, 1976). The reason for this is that the entity-relationship approach utilizes appropriate constructs, such as *entity, attribute relationship, degree of relationship, structural constraints and cardinality* (Flynn, 1998; Layzell and Loucopoulos, 1989).

The above-mentioned constructs have been identified in the light of the problem domain in Figure 3.3, since this figure identifies the entities responsible for causing flooding. Hence, the domain facilitates the identification of the interacting entities in the domain of flooding. There from, the concept of relational epistemology can be represented in Figure 3.4.

Figure 3.4 shows the actions between the interrelated entities, which are eventually used to identify the conceptual view of the model system as shown. It encapsulates the entities related to flooding in terms of intensifying factors, such as topography, meteorology and geological features along with river characteris-

66

tics. Figure 3.4 also encapsulates the entities that are subjected to the impact of flooding, such as route network, traffic flow, community, marketing and city. The relationships (as shown in Figure 3.4) between the entities have also been identified in the light of the problem domain of MCR. For example, the relationship type *'Flow'* involves the entities *'Route network'* and *'Traffic'*.

A relationship can also be termed as the interaction between interrelated entities Davies and Hossain, 2000). In the real world, this interaction between *'Route-network'* and *'Traffic'* could not be achieved without the flow of the traffic on the network. The encapsulation of this relationship is necessary to show the risk of route network flooding on the flow of traffic by the constructed phenomenology at the ontic level. Consequently, this could provide scope to visualize the impact of flooding on the characteristics of traffic flow in both qualitative and quantitative ways as pointed out in the illustration of the MCR problem domain.

In another example, the *'Impact'* relationship is participated by the entities *'Flood'* and *'Route network'*, because the impact of flooding on a route network cannot be recognized without its physical contact with floodwater. The encapsulation of this relationship is necessary in order to show the level of flood impact on a route network.

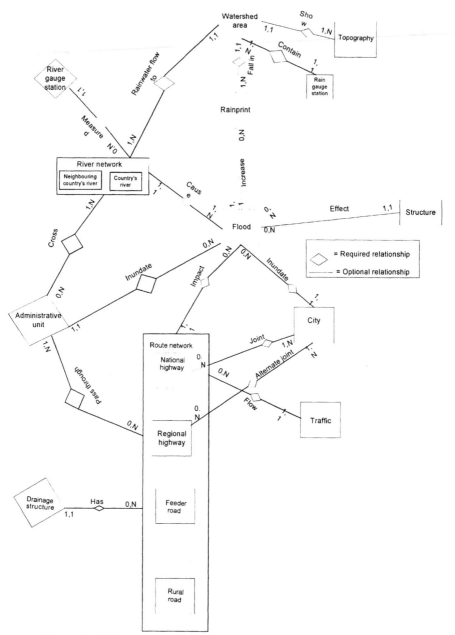

Figure 3.4: Conceptual View of the MCR to Assess Flood-Risk

68

Logical View of the MCR

As mentioned earlier, the ontic stage of the complete phenomenological overview of the Model of Computing Reality (MCR) must be constructed in a GIS environment since the problem is spatial in nature. There are many possible development platforms and languages that could be appropriate to develop a GIS. After a number of comparisons, usages, trials and analyses of the implementation strategies, results of the review clearly indicate that *ArcView* would be the most flexible and manageable development environment to develop such a GIS. It is important to note that the logical view of a computational model is guided by its implementation environment.

Since *ArcView* follows an object-oriented paradigm (ESRI, 2001), therefore, mapping of the conceptual view (developed by using ER model) into a logical view should be developed (using object-oriented model). It is necessary to represent such an approach in the schema of the *ArcView.*

Figure 3.5 illustrates the object-oriented logical view of the MCR. Each rectangular box in the figure represents an entity type or entity class. This feature represents the occurrence of a set of entity instances in an ER model. Each entity instance or class instance (occurrence of a class) is called an object. The straight line between two classes in Figure 3.5 shows that a one-to-one relationship exists between them.

A one-to-one relationship in an object model is termed an association. Similar to the class, a relational type represents the occurrence of a set of relationships. These relationship instances are called links in the object model (Flynn, 1998).

69

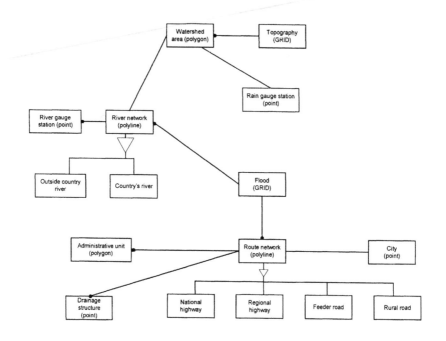

Figure 3.5: Logical View of the MCR

The one-to-one relationship type (association) between two classes implies that every object of one class should be linked to an object of the other class. For example, Figure 3.5 shows that the watershed area class has a one-to-one relationship with the river network class. A watershed consists of a set of sub-watersheds and a river network consists of a set of river links. Therefore, the one-to-one relationship between them implies that each river link should be associated with a sub-watershed.

The straight line with filled circle in Figure 3.5 between two classes indicates that a one-to-one relationship type exists between them. This implies that an object of one class may relate to many objects of the other class (class with attached full circle) in this type of relationship. For example, it can be seen that a one-to-many relations exist between the river network and the river gauge station. Such a relation type implies that a river link may relate to many river gauge stations.

The straight line with a triangle indicates that the lower-level classes are the subtype of the higher-level class. This is also called super type, as in the ER model. The relationship shown in the direction from the lower level to the higher level is termed as 'a kind of' relationship between the super type and subtype class.

This means that the subtype classes share some common attributes. For example, from Figure 3.5 it can be seen that the route network is an example of a super type class. On the other hand, national highway, regional highway, feeder road and rural road are examples of subtype class, as they are each a kind of route network. It can be concluded that all the entities and their relationships as they appear in Figure 3.4 are mapped into Figure 3.5, following the concept of object-oriented model. Hence, the relationship/interaction between the conceptual and logical views of the MCR can be understood.

PHYSICAL VIEW OF THE MCR

Physical view is concerned with how the objects and classes as illustrated in Figure 3.5 will be represented within the schema of the *ArcView*. The objects of the classes identified in Figure 3.5 can be categorized as objects with definite shape and those with indefinite boundaries. The watershed, river network, administrative unit, route network, city and rain gauge station class objects are examples of definite shape objects. On the other hand, topography and flood are examples of class objects with indefinite boundary.

Object and Data Structure in *ArcView*

In the *ArcView*, an object with definite shape is represented as a geometric object such as a point, polyline or polygon. The geometric objects are also called features. *ArcView* stores information on shape, location and attribute of each feature in three separate files and maintains an explicit one-to-one relationship between them. In this way, a structure to store information about a feature has been considered in *ArcView*. The structure represents a spatial data format, since it considers geometric location and attributes information of a feature. This spatial data format in *ArcView* is termed as '*ArcView Shapefile*', which is a combination of the three files. A *Shapefile* can be considered as a data structure for the definite shape objects, since it is associated with the features.

In *ArcView*, an object with indefinite boundary is represented as a GRID. The structure of a GRID constitutes an array of equally sized cells. These cells are assigned numeric/attribute values and arranged in rows and column. A GRID also consists of a number of files similar to the *Shapefile*.

Class in *ArcView*

An *ArcView Shapefile* stores information on the features/objects of the same shape. Therefore, it provides scope to store the information on all the features associated with a class. For example, information on all the river links of the river network feature class can be stored in a *Shapefile*. Thus a *Shapefile* can be used to represent the information of a class. The class can also be termed a feature class (Longley et. al., 2001), since it is associated with features. A *Shapefile* is visualised as a feature theme in *ArcView*. Thus, a feature theme can be used to represent a class in the schema of the *ArcView*.

A feature theme in the *ArcView* performs the following tasks:

- draws the shape of the feature in a view;
- links the attribute information with the shape;
- provides a means of visualizing the attribute information as a table named *'Feature Table'*.

A *Feature Table* visualizes each feature of a theme as a record and each attribute as a field. A feature table should contain a field named *'shape'* through which it maintains links with the shape in the view. Thus, it can be seen that a feature theme in the *ArcView* can be used to visualize a class of definite shape objects and their geometric and attribute data in a table. The visualization of the GRID in the *ArcView* is called the GRID theme and can be used to represent a class of indefinite boundary objects.

Representation of Classes and their Objects

By considering the object and class representation in the schema of *ArcView*, the river network and route network class can be represented as a linear feature theme and their instances/objects as polylines along with their locations and attribute information. Figure 3.5 considers two types of river networks. They can be visualized either as a separate theme or under the same river network theme. In the same way, various types of route network can be represented in the schema of *ArcView*.

The city, rain gauge station, river gauge station, and drainage structure classes can be represented as a point feature theme and their instances/objects as points. The watershed and administrative unit classes can be represented as a polygon feature theme and their instances/objects as polygons. Flood and topography classes can be represented as a GRID theme and their instances as GRID, as they do not have a definite boundary.

Thus, from the above, it can be seen that each class identified in the object model as well as its associated objects can be represented in the schema of the *ArcView*. Hence a mapping from logical view to physical view of the MCR is achieved.

Relationship Representation in the *ArcView*

There exits a one-to-many relationship between the topography class and the watershed area class. This implies that an instance of topography class can be linked to many instances of watershed, which are sub-watershed areas. In *ArcView*, the topography can be represented as a GRID theme. Therefore, only one GRID theme instance or object needs to be related to all the sub-watershed areas of the watershed class.

The relationship type between the flood and river network is one to many. This implies that the generation of a flood is related to many instances of river network class, which are river links. This will also be shown in the next section, which will demonstrate the generation of a flood from the data associated with the river links. The flood will be visualized as a GRID theme.

In the same way, the relationships between the other classes can be understood and the establishment of the relationship type between the classes will be demonstrated in later sections. Hence, the mapping of the relationship from logical to physical level of the MCR can be understood.

Thus, from the above explanation it can be concluded that a linkage among the three views of the MCR is achieved. This enables the construction of an appropriate phenomenology for MCR. This formulation is next used to analyze the problem of flooding from multivariate perspectives. We will show this in the next section while connecting with the concept of circular causation and the IIE-process model.

APPLICATION OF THE MCR

This section presents the results generated by using the Model of Computing Reality for Bangladesh as a case study to apply the MCR, since flooding is a perennial problem in this country. Figure 3.6 illustrates the generation of flood scenario for a district of Bangladesh with specific amounts of rainfall. It also visualizes flooding in the area in a three-dimensional way since this enables the analysis of flooding from various perspectives.

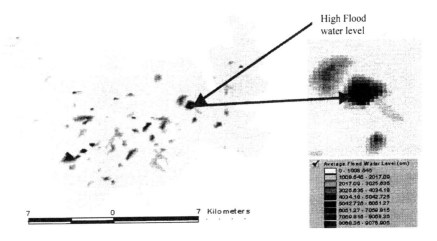

Figure 3.6: 2-D Flood Generation

Figure 3.7: Danger Point Locations

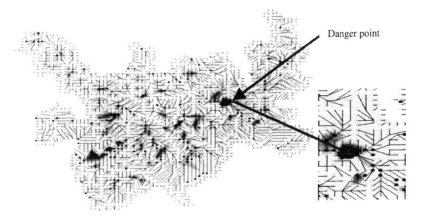

75

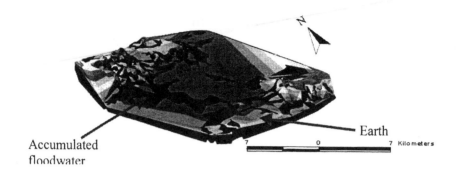

Accumulated
floodwater

Earth

7   0   7 Kilometers

Figure 3.8: 3-D Flood Generation

Figures 3.6 and 3.7 demonstrate the risk of flooding, while Figure 3.8 illustrates it in a three-dimensional way, enabling the identification of the location of flood-risk areas. Here, rainfall has been considered as the flood intensifying factor along with other variables as depicted in Figure 3.4.

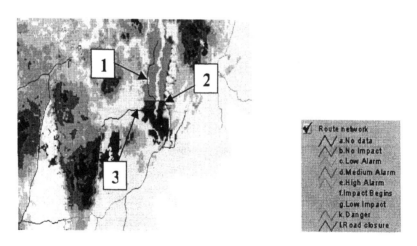

Route network
a.No data
b.No Impact
c.Low Alarm
d.Medium Alarm
e.High Alarm
f.Impact Begins
g.Low Impact
k.Danger
l.Road closure

Figure 3.9: 50-Year Flood-Risk on Route Network

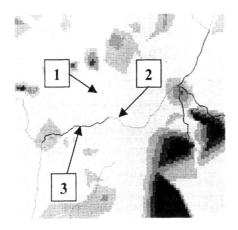
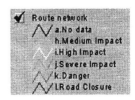

Figure 3.10: Artificial Flood Risk on the same Route Network

Figures 3.9 and 3.10 visualize and classify the risk of flooding on the route network for a 50-year flood and an artificial flooding in an area. Figure 3.9 illustrates that the route links 1, 2 and 3 are completely closed because of the 50-year flood cycle. On the other hand, Figure 3.10 illustrates that artificial flooding can produce less flood-risk on the two links [1, 2], while it produces the highest risk on the third link [3]. Thus, the interaction between flood and the route network can be perceived. In this way, different scenarios of flood risk on the same route link for different kinds of flood can be understood.

Hence, the MCR allows testing of the role of different policy-induced variables on the reduction of flood risk on objects by way of circular causation. For example, by increasing water discharging capacity of a river, flood-risk on a link can be reduced. This can be visualized using the MCR. In this way, the MCR allows the study of the effects of various policy variables. Such variables are chosen by a systemic participatory process, as shown in Figure 3.3 on the reduction of flood-risk on objects.

The effect of 'Road Level Modification' policy variable on the reduction of flood-risk/impact is shown in Figure 3.11. This will in turn help to ensure the movement of vehicles on the route network during flooding period in an area, and will thereby safeguard socioeconomic activities. Hence, the end result is the evaluation of the wellbeing function as a consequence of the policy induced variable. In the same way, the effect of other policy variables, like 'installation of culvert' on the route network to reduce the risk of flooding, can be shown using the model. 'Alternative path selection' can be considered as another decision variable to be used to ensure the traffic movement during flooding period, and thereby, to reduce the risk of flooding on the movement of traffic. Thus, it can be seen that the MCR supports the decision-making process of the selection of various policy variables to reduce the risk of flooding in a circular causation way.

CONCLUSION

In this chapter the methodology of relational epistemology has been used to develop a Model of Computing Reality (MCR) to assess the risk of flooding on various entities. It is shown that the methodology of relational epistemology can be an appropriate methodology for developing the MCR, since it enables the integration of the three views of the MCR by way of circular causation. This is evident from the development of risk assessment framework in the light of relational epistemology, which allows the diversification of risk and its evaluation in terms of a socioeconomic wellbeing function.

The model development framework is also guided by the three views (conceptual, logical and physical) of the relational epistemology. Hence the development framework system can produce reliable and dependable results to support decision-making process on flood-risk reduction. Finally, the application of the MCR has been demonstrated. The result shows that the IIE-process model enables a process of relational reasoning for reaching a decision to reduce flood risk by analyzing different scenarios of policy variables in a way of circulation causation within a computational model.

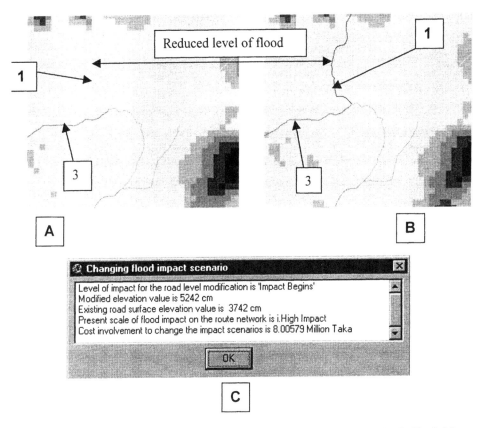

Figure 3.11: Application of 'Road Level Modification Policy' Variables [Ontic Stage in the Complete Phenomenological Model of MCR]

REFERENCES

Burch, J. G. 1992. *Systems analysis, design and implementation,* Boston, USA: Boyd & Fraser Publishing Company.

Chen, P.P., 1976. "The Entity-Relationship Model – Toward a Universal View of Data", *ACM Transactions on Database Systems*, Vol.1, No. 1, pp. 9-36.

Choudhury, M.A. and Hossain, M.S., 2006. *Development Planning in the Sultanate of Oman,* Edwin Mellen Press.

CRED, 1993. *EM-DAT (a disaster event database),* Centre for Research on the Epideminology of Disasters, Catholic University of Louvain, School of Public Health, Clos Chapelle aux champs 30-34, 1200 Brussels, Belgium.

Cutts, G., 1991. *Structured Systems Analysis and Design Methodology* (2nd ed), Oxford, Eng: Blackwell.

Davies, C.G. and Layzell, P.J.,1993. *The Jackson Approach to System Development: an Introduction,* Bromley, UK: Chartwell-Bratt.

Davies, C.G., and Hossain, M.S., 2001. "A System for the Simulation and Prediction of Floods".In Kidner, D.B. and Higgs (eds.), *GIS Research in the UK, Proceedings of the GIS Research UK 9th Annual Conference GISRUK 2001,* April 2001, University of Glamorgan, Wales, UK, pp.490-494.

ESRI, 2001a, *ArcView GIS Literature,* Environmental Systems Research Institute, California, USA. http://www.esri.com/library/whitepapers/.av_lithtml

Flynn, D.J., 1998. *Information Systems Requirements: Determination & Analysis,* McGraw-Hill, London.

Flynn, D.J. and Fragoso Diaz, O., 1996. *Information Modelling: An International perspective,* Prentice-Hall, London.

Hossain, M.S., 2002. *A Method to Reduce Flood Impact on Road Transportation Systems,* PhD. Thesis, Department of Computation, University of Manchester Institute of Science and Technology (UMIST), U.K.

Hossain, M.S., 2000. *An Investigation into the Provision of a GIS to Assess Risks Associated with Flood Disasters on Road Transportation,* MPhil Thesis, Department of Computation, University of Manchester Institute of Science and Technology (UMIST), U.K.

Hossain, M.S., 2005. *Computer Modelling,* Lecture Note, Institute of Islamic Economics and Finance, Trisakti University, Jakarta, Indonesia.

Hossain, M.S., 2005. Developing Computer-Based Information System to Simulate Money and Real Economy Linkage. In Choudhury, M.A. (eds), *Money and Real Economy:Studies in Real Money linkages with Social Issues, Economic Transformation, Institutions and Markets*, Leeds, UK: Wisdom House, pp. 66-80.

Jeffrey, A. H., Georg, J.F. and Valacich, J.S., 1999. *Modern System Analysis and Design* (2nd ed.), Addison Wesley, Reading, USA.

Kendall, K.E. and Kendall, J.E., 1988. *Systems Analysis and Design*, London, UK: Prentice-Hall, Inc..

Kendall, P.A., 1996. *Introduction to System Analysis & Design: A Structured Approach* (3rd ed). London, UK: Irwin.

Layzell, P.J. and Loucopoulos P., 1989. *System Analysis and Development* (3rd ed), Bromley, U.K: Chartwell-Bratt, Bromely.

Longley, P.A., Goodchild, M.F., Maguire, D.J. and Rhind, D.W., 2001. *Geographic Information Systems and Science*, John Wiley & Sons Inc., New York.

Loucopoulos, P. and Karakostas, V., 2002. *Requirement Engineering*, Department of Computation, University of Manchester Institute of Science and Technology.

Russel, S. and Norvig, P., 2002. *Artificial Intelligence*, Pearson Education, Inc.

Rumaugh, J., Blaha, M., Premerlani, W., Eddy, F. and Lorensen, W., 1991. *Object-oriented Modelling and Design*, Prentice-Hall Inc., Englewood Cliffs, New Jersy.

Turban, E. and Aronson, J.E., 2001. *Decision Support Systems and Intelligent Systems*, Prentice Hall International Inc.

Turner, W.S., Langerhorst, R.P., Hice, G.F., Eilers, H.B. and Vijttenbroek, A. A., 1987. *System Development Methodology*. Netherlands and North Holland, Amsterdam, Netherlands.

CHAPTER 4: CONTRASTING MEANINGS OF REALITY IN SYSTEMS FRAMEWORK

Computing Reality is a project on rethinking what are we to compute as the essential observed truth in terms of its application to various issues and problems of world-systems. In earlier chapters we have examined that truth can be elusive even in the best of scientific reasoning, what is often called the Scientific Method. There is also the danger that a non-substantive understanding of the true essence of what science means and thus what the scientific method ought to be, relinquishes the quest for truth to a mechanistic understanding of methods. Thus the danger of socio-scientific reasoning devolves into misinterpretation of the essence as also of the understanding of the true relational nature of world-systems that science wants to construct and interpret. Subsequently, the mechanistic interpretation of the world-systems and their system-entities runs the profound danger of a misconstrued direction in scientific reasoning in questing for truth rather than mechanism qua temporality in the realm of essence.

The questioning on the essential nature of scientific inquiry in quest for truth and its subsequent applications can be a project of postmodernism. Yet there are serious matters embodying postmodernism on ethical and moral issues of scientific methodology that must be critically examined. A number of important groundwork thus opens up in the quest for reality in the sense of searching and discovering the essential nature of truth. Such a project in the ethical dimensions of scientific methodology (Bohr, 1985; Einstein, undated) is next followed by formalism and application premised on the concept and essence of truth. This latter perspective is central in the formalism of neurocybernetic and system theory as the project of formalizing and applying the conceptual worldview of unity of knowledge to the manifest complex phenomena.

We characterized the total project of epistemology leading to the conceptual stage of ontological formalism and its jointly premised consequences in the ontic stage of materiality as the comprehensive meaning of phenomenology of unity of knowledge. The combination of these three stages of the phenomenology of unity of knowledge is uniquely and universally premised in all world-systems and their system-entities. We have, though not formally established, the methodology and model of phenomenology of unity of knowledge in the alternative groundwork of unity of knowledge beyond just explaining the nature of such a phenomenological project in unity of knowledge and the world-systems, and inquiring some recent contributions in this field in the area of neurocybernetic and system theory. The combination of the two perspectives of the comprehensive meaning of scientific essence, namely the precept of reality embodied in the epistemological, ontological and ontic contents and the application of this to neurocybernetic and system theory in unity of knowledge, is the project of *Computing Reality*.

OBJECTIVE

The objective of this chapter is to further expand our inquiry on the project of phenomenology of unity of knowledge of neurocybernetic and system theory in the light of contemporary studies on phenomenology in general. We will address the questions: How is the epistemological, ontological and ontic continuity of our phenomenological theory of unified world-systems contrasted or compared with the classical and contemporary theories of the same? The second question is to critically examine whether such other theories of phenomenology lead to the development of the comprehensive formalism of phenomenology of unity of knowledge as we perceive it in our understanding of computing reality. To begin our purported investigation in this chapter we will commence with the emergent interest in postmodernism that has been targeted as a new epistemological way of looking at old issues and problems of world-system studies. This will be followed by a discussion of postmodernism within a phenomenological context. The chapter then moves on to an examination of liberalism and the phenomenology of unity of knowledge as contrasting paradigms in the study of learning systems, complexity versus linear reasoning, markets, institutions and theory of value.

THE POSTMODERNIST WORLD-SYSTEM QUANDARY

Fitzpatrick (2003) cogently summarizes the worldview of postmodernist inquiry. A number of characteristics of postmodernist project are provided. The characterization of postmodernism revolves around the delineation of world-systems as constructions out of languages and discoursed descriptions made possible by human perceptions and reasoning. This is the Dasein referred to by Heidegger (1988). Consequently, the postmodernists argue that there is no uniqueness and universality of meta-knowledge provided from outside the matters being discoursed. Knowledge pertains only to the discourse itself, and this is not convergent. The idea of a unique and universal knowledge premise and the continued reference of the nature of reality in the experiential domains to such a pre-determined epistemology taken together, are referred to as meta-narrative. The postmodernist school abandons this purview. In this regard Lyotard (1984, p. 35) writes: "In contemporary society and culture – post-industrial society, postmodern culture – the question of legitimation of knowledge is formulated in different terms. The grand narrative has lost its credibility – regardless of whether it is a speculative narrative or a narrative of emancipation."

The postmodernist analytical inquiry thereby, reduces to non-foundationalism. Thus a recursive process of constructed anti-thesis between concepts and forces prevails across both the socio-political and natural science

85

domains. Consequently, the characteristic of the postmodernist world-system is based on non-convergent circular argumentation over issues and problems under discourse. Fitzpatrick (op cit, p. 128) writes on such an unsettled dynamics of postmodernist reasoning in the following words: "Ours then is an age of simulations that endlessly refer only to other simulations. The infinite circularity of these self-references is what Baudrillard calls the *simulacra*: everything is a reproduction of other reproductions." Besides, postmodernism argues in favor of a system of capitalism that while it competes with other systems also competes within itself by the process of its own fragmentation. This is of the nature of Darwinian mutations in socio-scientific systems that yield to perpetual individuation of the entities. This is a characteristic of liberalism as a world-system and of its system-entities existing as variables, relations and sub-systems.

According to Wallerstein (1998) the postmodernist ambition is to study social reality in the large scale socio-political universe. This is Wallerstein's Eternal TimeSpace. Postmodernism abandons operating over incremental changes. This notion contrarily is found in the economic argumentation of neoclassical economic theory, wherein time is non-existent as an organism. Time is simply implied as datum. Wallerstein refers to the incremental nature of time as Episodic Geopolitical TimeSpace. The totality of these two TimeSpaces is thus contrasted between two ends. In light of such contrasts the capitalist world-system as a reflection of the liberal paradigm enforces on the one hand a convergence and its mutation. On the other hand, the postmodernist thought premised on large scale change and structural transformation of the capitalist order of production, consumption, governance and progress, views the coming world-system as a rejection of the capitalist mindspace.

This contrast is as untenable in Wallerstein's thesis of the internally opposed timal nature of social reality as it is self-contradicted in postmodernism.

This contradiction in the two paradigms is seen in terms of the postmodernist attitude to reject the convergence hypothesis into global capitalism, a convergence doctrinaire otherwise emphasized by Fukuyama (1992). With Wallerstein the contradiction of ideas is caused by the inability to reconcile these opposite TimeSpace concepts with his yearning for a unified epistemology of the world of knowledge. In fact, in Wallerstein's world-system with competing mutations along the path of historical evolutions, there exists only complexity qua 'chaos' and entropy. This is an entropy model of complex world-systems with unstable evolutionary equilibriums. No unity of knowledge for a constructive structure of the world-system can be formed in such a social loci of multitudinous social forms. Permanent and perpetual complexity out of chaos abides. In regard to the cyclo-ideological TimeSpace argument provided by Wallerstein on the nature of the capitalist world-system, which is a property represented equally by the inherent liberal paradigm, he writes (Wallerstein, p. 57):

> For social science, the rise of complexity studies represents an epistemological revolution. On the one hand, it undermines totally the basis of the concept of eternal TimeSpace, while at the same time rejecting that of episodic geopolitical TimeSpace, substituting for it the rules of social processes for as long as these rules are relevant. For the 'orders' that are represented by these rules constantly yield place to periods and 'loci' of 'chaos' out of which new 'orders' are constantly regenerated. This is precisely the concept of structural TimeSpace with cyclico-ideological TimeSpace located within it ......

We argue here that since postmodernism represents anomie in its dialectical process of perpetual conflict and evolution across Wallerstein's kinds of eternal TimeSpace notion of history, it carries a methodology that is not tenable in the perspective of the phenomenology of unity of knowledge that we are aspiring to develop and explain. The contrary dialectical and neo-liberal paradigms mentioned above deny all the central tenets of the phenomenology of unity of knowledge, upon which we aim at constructing the nature of computing reality. The central status of the phenomenology of unity of knowledge logically rejects or explains the methodological impossibility for the existence of a unique and universal world-system paradigm premised in any other worldview. The same precepts are extended to embedded systems and their system-entities and their evolutionary unifying circular causal relations. We have discussed these ideas to an extent in earlier chapters when introducing, but not completing, the formalism of the emergent methodology and methods followed by their applications to verities of neurocybernetic and system theory.

INABILITY OF THE LIBERAL PARADIGM AND GLOBAL POLITICAL ECONOMY TO ADDRESS THE SYSTEMIC WORLDVIEW OF UNITY OF KNOWLEDGE

One of the paradigms where unity of knowledge in the neurocybernetic and systems sense becomes impossible is liberalism. The socio-scientific reasoning inherited from the liberal and political economy standpoints of concepts and applications also suffers from the problem of their impossibility. The paradigms that are today gripping the world-systems in the order of intellectual thought and global governance are manifestations of the liberal failure. It has also been prototyped in the field of 'global political economy'.

To evaluate these claims we first need to understand the contexts of liberalism and its reflection in the current thinking on global political economy. We need also to understand how these concepts and the attenuating governance formed by liberalism (Minogue, 1963; Weaver, 1977) are overshadowing the world in the name of capitalist globalization and its postmodernist questioning by an equally untenable project of global political economy (Palan, 2000).

Characterizing Liberalism

Minogue explains the liberal mind as one of peace that endeavors to organize human actions around social consensus over contentious issues. It premises such consensus on the ideals of the rational individual framed on individuality as the ideal to be reasonable. Individuality itself is a bundle of egos but at the same time linearly combined in human personality and reasoned will. Under this kind of the portrait of self the liberal agenda in social, economic and political fields, liberalism extends individuality to the social whole. Within this same kind of linearly reasoned construction of the socioeconomic and political order, precepts such as of happiness, justice, liberty and freedom take shape and are applied. The implications of such methodological individualism are thereby projected thoroughly all through the social and political constructive mechanism. Some examples can be taken up here.

The ultimate parting knell of liberalism rests upon its claim on materiality, abandoning the metaphysical relevance of ethics and morality. This concern has always been the basis of contest between the Christian Church and liberalism as a philosophy of life. The idea of material reductionism is explained well by Minogue (op cit, p.167, edited): "It (materialism) has come to describe a life devoted to the pursuit of material objects and advantages. ... The Christian Churches have attacked this kind of materialism on two grounds: Partly the Church saw liberalism to be ruthless and uncharitable, a selfish trampling over the interests of the powerless; and partly it grounded liberalism in a way of life which refuses to take seriously the religious mystery of the universe."

The concepts of happiness, freedom, liberty, and thereby human personality are formulated in the framework of materialism. The liberal dictate announces these other attributes to be material objects of pursuit. Thereby, happiness is relegated to the acquisition of human artifacts for the pleasure of self.

Even when social justice as attribute of human happiness is pursued, it means the possibility for acquiring material benefits from the market order through the free and untrammeled process of economic competition. Economic liberalism treats the individual personality as a being competing for scarce resources for material satisfaction within the sanction of the natural law of liberty. The natural law sees both man and nature to be free in creation. This kind of the law of liberty is effectively manifested in Smith's words, despite that Adam Smith wrote remarkably on human sentiments in his Theory of Moral Sentiments (1937).

Smith (1937 [1776], p. 14) wrote:

> It is not from the benevolence of the butcher, the brewer, or the baker, that we expect our dinner, but from their regard to their own self-interest. We address ourselves not to their humanity but to their self-love, and never talk to them of our own necessities, but of their advantages. Nobody but a beggar chooses to depend chiefly upon the benevolence of his fellow citizens.

Yet in his Moral Sentiments Smith (1996, [1753], p. 124-125) wrote:

> All the members of human society stand in need of each other's assistance, and are likewise exposed to mutual injuries. Where the necessary assistance is reciprocally afforded from love, from gratitude, from friendship, and esteem, the society flourishes and is happy. All the different members of it are bound together by the agreeable bands of love and affection, and are, as it were, drawn to one common centre of mutual good offices.

Competition versus Participation as a Methodological Concept

The contrasting notes in the above two Smithian perceptions of human nature reflect how human empathy is perceived in the realms of moral values and market processes. The principle of natural liberty is copied from the state of nature as an entity free of intervention once ordained by God, and is applied to the human order, as in the case of the market process. The differing tilt from the Smithian thought in The Moral Sentiments to the Wealth of Nations is seen to disengage the moral from the material orders. The material order of market competition reverses the moral sentiments of human cooperation.

Yet the economists keep on interpreting Smithian competition as cooperation (Parkin & Bade on Lindbeck Assar's views on cooperation and competition, 1991). The argument presented is that by assuming natural liberty to govern the preferences and behavior of a freely competing individual in market exchange every individual likewise adjusts to an auctioneering process that the invisible hand of market exchange establishes automatically between buyers and sellers taken up in their various categories, in order to acquire mutual benefit. The misconception here is on the understanding of the substantive concepts of competition versus cooperation taken up in their methodological sense.

Social and economic competition is premised on the assumptions of scarcity of resources and the prevalence of full information or partial information in exchange. These assumptions in turn generate the ego in self to obtain maximum benefits from the market process of exchange under the assumption of scarcity and the quest for full information. Even in the case of less than full information in market exchange the assumption of scarcity results in competition for economic surpluses as opposed to normal profits of perfectly competitive market. The assumption of resource scarcity and the quest for information together form the basis of economic rationality of the self-interested individual. The theory of perfect competition views the self-interested individual to methodologically apply the principle of natural liberty in the market process. The individual thereby competes with another under the guise of resource scarcity, quest for full information, and the sanction of the law of natural liberty taken up in the social context.

Competition, resource scarcity and the quest for full information to maximize individual gains through market exchange are thus coterminous properties of the rational economic agent. Economic theory assumes without explanation as to how the maximum gains are attainable through the market process (Shackle, 1972). If and when such a state of exchange is attained, then methodologically speaking, no further scope of learning remains between the agents in exchange. The exchange system has to be rejuvenated by external shocks to revive a new system of resource allocation into competing ends that are once again governed by the same structural assumptions of economic rationality. Such new states do not arise from a learning process embedded in market exchange.

Contrarily, such a learning experience otherwise can be systemically endogenized into market exchange. The result would then be dynamic regeneration of organically perceived social processes. These would manifest themselves in the theatre of mutually attained, ratified and creatively evolved social contracts. Markets are then seen as systems of social contracts (Choudhury, 1996) rather than methodologically formulated exchanges between ultimately optimized states leading to non-learning agent behavior.

The meaning of competition in the liberal economic, social and political contexts is therefore a methodological implication of optimized and steady-state equilibrium exchange processes that result in non-learning states. The same idea of competition is extended to the theory of institutionalism in liberal paradigm by the linear aggregation of agent-specific preferences, utilities and benefits, in spite of ethical induction according to the liberal perception (Hammond, 1987). Here the institutional formation of preferences is grounded on the individualistic nature of self and markets (Buchanan, 1999). The benefits are grounded on the material perception of liberalism. Included in such benefits are the materialistic origins of the concepts of happiness, freedom and liberty. Happiness is based on the acquisition of bundles of benefit-maximizing commodities. Freedom is based on the untrammeled position of self in determining social contracts. Liberty means extension of the natural liberty law into social contracts with their protection by the institutional system that liberalism promotes in quest of establishing its way of life and thought extensively.

An example of methodological competition and the liberal idea of social participation can be idealized here by a story. A Dog and a Fox cross paths in a forest and enter into a dialogue. The Dog boasts of how well his master takes care of him in so many ways. The Fox hears silently and then asks the Dog, "What is the mark around your neck, cousin Dog?" Dog says that it is the mark left by the chain with which his master ties him in the manger. Fox says, "Oh! Even with all that you get from your master, you still do not have that which is the most precious to me – my freedom from bondage in the midst of abundant nature". The Dog is subservient to the social contract of the market process in the midst of competition between winners and happy losers. The Fox is an inhabitant of the realm of possibilities yielded by his own productive participation in the abundance of resources that nature holds out for him.

The lesson here is that resource scarcity is a methodological convenience of economic theory and of social contract in market exchange. The same social contract is then translated into the behavior of individuals and organizations extensively. Participation as opposed to liberal market exchange is the medium through which newer resource possibilities and behavioral changes are continuously generated in consumption, production and distribution of the gains obtained from markets through participation and productive activities. Such assumptions are molded into behavior by the powerful governance of liberal world-system over individuals and institutions globally.

Cooperation *qua* Participation in Market Exchange

Cooperation belongs to the category of human sentiments. It is premised on sharing and organic participation. Methodologically, the concept of sharing is translated as participation between human agents and between human agency and the socio-scientific domain. In the foregoing discussion on competition and the participatory alternative of markets as social contracts, human empathy assumes the role of endogenous values defining preferences and material acquisitions. Consequently, such an endogenous role is carried through the discursive process of human society into the organization of preferences, the determination of commodity bundles and the organization of institutions along directions supporting the endogenous preference and commodity determination. The result is endogenous reformation of preferences that enter market exchange and the nature of the exchangeable. We invoke here our concepts of social and spiritual capital that were given in earlier chapters while explaining ethically induced commodity bundles.

In the presence of human empathy qua sharing and participation there is automatically a discursive mechanism that underlies the process of preference formation and determination of commodity bundles. Sen (1999) refers to such developmental artifacts taken together as the 'capabilities for social functioning'.

In Sen's case the three critical attributes of capabilities to function are sustenance, self-esteem and freedom. The underlying idea is this. While eating of fruits gives you pleasure, this enjoyment comes out of the essence of the fruits and the individual's disposition for the fruits. Malaysians, Thais and Indonesians enjoy the fruit called Dorian as the King of Fruits. Yet Dorian can be the most avoided fruit for others because of its strong smell. Therefore, agricultural development in Dorian cultivation would be promoted as a commodity bundle because of the intrinsic value it carries, not simply from sales but also from the perception and preferences of consumers. These are entrenched in such a commodity bundle and in the agricultural development process. Lancaster (1966) referred to the organic understanding of commodity with its embedded values in it as attributes qualifying the demand function.

The embedding of such values in commodity bundles defining the development process, as in the agricultural sector, is extant in the entire understanding of the development process. By a conscious understanding of such embedded values the organic nature of demand and supply of commodities results in non-linear and complex demand relations. These cannot be conceptualized and estimated methodologically by the same kinds of market exchange equilibrium and benefit-maximization concepts as harbored by the concept of competition in the framework of market mechanism.

Contrarily, the embodiment of human values, such as empathy and cooperation, carries with it at every moment of a decision-making and allocations, the presence of economies of scale and ethical enhancements of consumer preferences. A learning system of exchange is thereby generated in the market mechanism. It causes permanent perturbations and evolutionary equilibriums due to simulation of every value-induced transformation of preferences and determination of commodity bundles that get induced by ethical values. We can think of such commodity bundles, like in the case of social and spiritual capital, as endogenously value-induced material artifacts, and the market exchange as the evolutionary social contract that forms such ethically induced material artifacts.

Markets now exist in the medium of continuous learning by moral and ethical values contributed by the principle of human empathy qua sharing and participation. The ethical market transformation process is a deeply dynamic and creative process caused by the endogenous nature of resource allocation, as opposed to exogenously driven re-distributions and resource injection into the economic system by virtue of exogenous policies, institutions and governance. Morality, ethics and values ought to be inculcated in the human psyche, rather than be imposed consequences. Markets as social contracts for the endogenous realization and regeneration of morality, ethics and values depend on proper guidance by laws, rules and discursive relations.

The emanating theory of markets as social contracts, is thereby deeply epistemological, ontological and ontic in substance. The system consciousness of the neurocybernetic and system theory now connects well with such a knowledge-induced market transformation process. Within this kind of systemic understanding of market transformation with knowledge-induced preferences and commodity determinations the learning process gets deeply embedded.

THE CONTRAST BETWEEN METHODOLOGICAL COMPETITION AND COOPERATION IN EXTENDED SOCIO-SCIENTIFIC PURVIEW

The difference between the methodological and behavioral concepts of competition and cooperation on social matters is now evident. Economic competition in all cases of market conditions means the particular behavior of individuals as buyers and sellers, or likewise of similar agents with their exchangeable. Such behavior arise from the underlying conditions of resource scarcity, quest for full information and cessation of learning by interaction in the final state of steady-state equilibrium and optimum allocation of resources among competing ends. Social and economic cooperation is the participatory nature of discourse for learning and sharing between exchange agents in concert with the nature of ethical markets and guidance by means of moral laws, rules and policies that are amenable to learning preferences. These are the endogenous preferences. In this context both the state as well as the policy (control) variables becomes endogenous. Preferences by learning become dynamic, along lines of such participatory learning. (Intrilligator, 1971; Vanek, 1971).

Because the concept of competition and cooperation respectively spans over domains of state and control variables, behavioral and engineering perspectives, they become differentiating concepts for the entire socio-scientific order. It is therefore important to note the meaning of organic participation, not simply at the human and institutional levels but also within the scheme of things and between the human and non-human worlds. The fundamental assumption underlying such a generalization in such an extended concept of participatoriness is that both the human and the non-human worlds learn by the same kind of dynamics involving epistemology, ontology and ontic forms. All that matters is that the mind receives such a learning impact from the real world and responds in terms of understanding those unifying processes of the real world by means of the unique and universal worldview of unity of knowledge. This worldview emanates from the IIE-process perceptive in the midst of the epistemological, ontological and ontic phases within the comprehensive phenomenological model of unity of knowledge.

We will discuss later on with respect to learning system dynamics that, under conditions of participation, resources get continuously enhanced. New production and risk diversification possibilities are opened up for participatory shareholders and stakeholders. The socio-scientific systems, with which the mind interfaces, unravel the methodological possibilities. Consequently, only dynamic preferences prevail and evolve in such learning environments. Steady-state equilibriums become stable evolutionary equilibriums by the force of learning (knowledge-induced). Optimization goals of objective functions in the socioeconomic context are replaced methodologically by simulation objective criteria with learning constraints.

These are substantive methodological issues that involve evolutionary dynamics relating to the epistemology, ontology and ontic consequences of learning systems. The convergent methodology for such understanding of socioeconomic issues is entrenched in the Interactive, Integrative and Evolutionary Process, i.e. the IIE process-oriented model that we have introduced before. It will be taken up in further methodological depth and applied to various issues and problems as we proceed on.

Ethical Implications of Participatory Extensions in Socio-Scientific System

The emergence of the participatory dynamics in ethical transformation of market theory is a substantive theory spanning across several dimensions. In the Technical Appendix we explain such a dimensional extension and systemic interaction between system-entities and the human domain, concerning the endogenously learning market transformation by knowledge induction.

The idea of participation when extended to the non-human domain can be subsumed in the generalized principle of pervasive complementarities. We have discussed this principle earlier. It will be taken up in depth later on as well. The attribute of complementarities arises from the paired nature of creation. That is, all things in the universe are created in pairs. In core scientific terms even the number system, which is a purely mathematical precept, is a paired system of real and imaginary numbers. All such entities co-exist to perform a given function of systemic unity conveyed by meaning and solvability.

Pairing in the socio-scientific sense thus conveys two coterminous meanings. Firstly, pairs are between relational system-entities. Examples are of two systems, one comprising men and the other women; the baby and mother are parentally most united together in the animate worlds; fire and heat coexist, etc. These are paired by their complementary functions in accomplishing a social task. Secondly, pairing is between diverse systems. Consider the statements, 'birds of the same flock together'; salt water and fresh water do not mix; positive and negative charges co-exist in the atom.

In the first case a complete social function is accomplished by the entities unifying in the light of the rule of order, equilibrium and harmony between learning system-entities. If there is no physiological need according to social accepted laws there would be no family. Hence there would not be strong bond between parents and their children. Families break up and children belong to broken homes. Instead, when mother and children establish the parental bond then a social environment prevails.

Likewise, if fire does not generate heat the constitution of fire changes into other unknown things. Conversely, if heat does not result in fire, then there would never be forest fires. There remains the permanent objective function for all pairs to attain a sense of mutual harmony of purpose to generate balance. Such harmony is established by constructive learning between the system-entities on the framework of laws, rules, ethical conduct and moral values.

In the second case, the diverse systems remain internally consistent in terms of co-existing relations between the system-entities. But the systems do not interact in the sense of changing their individual nature through such interaction. The social function is to maintain harmony between systems. For example, there ought not to be ecological deprivation to alter the proportions of salt and fresh water in given environment needs. There ought not to be destruction of bird habitats that would create imbalance in the diversity of bird population for sustaining ecological diversity. There ought not to be such technological ventures that would tamper with the well-established sub-atomic relations. In every case we note that complementarities as pairing between diverse system-entities within and across systems must be maintained by not enacting such human behavior, technological change and ecological interruptions that tamper with the laws of nature.

The learning process between intra- and inter-systemic entities is preserved and enhanced by the circular causation relations between system-entities in the environment of unity between them according to the law of science and social order. As all such system-entities are children of a balanced universe in the framework of learning through circular causation *inter*relationships between them so as to maintain the universal law. Humans learn about observed system-entities and their domains in all of socio-scientific order only in the midst of an organic experience of concepts, discourse, participation, guidance and inferences. The same process of learning is then continued in evolutionary creativity. All these are intrinsic in the Interactive, Integrative and Evolutionary Process (IIE-process) that we have introduced earlier, pending deeper formulation in subsequent chapters.

The organic medium of learning in unity of knowledge can be explained in the way as Thomas Kuhn explained the learning experience of a child. Since both the animate and inanimate world-systems can be thought of as being embryonically determined by primitive concepts, therefore Kuhn's example conveys a pervasive meaning of neurocybernetic and learning system. Every world-system in the socio-scientific worldview of unity of knowledge shares in the unique and universal construct of the unitary episteme and carries it forward through the ontological principle into ontic reality of quantification, measurement, application and inference. The scientific theories do not escape this kind of generic principle.

Kuhn (1970, reprint, p. 190) writes:

The child who transfers the word 'mama' from all humans to all females and then to his mother is not just learning what 'mama' means or who his mother is. Simultaneously he is learning some of the differences between males and females as well as something about the ways in which all but one female will behave toward him. His reactions, expectations, and beliefs – indeed, much of his perceived world – change accordingly. By the same token, the Copernicans who denied its traditional title 'planet' to the sun were not only learning what 'planet' meant or what the sun was. Instead, they were changing the meaning of 'planet' so that it could continue to make useful distinctions in a world where all celestial bodies, not just the sun, were seen differently from the way they had been seen before.

100

Einstein (Commins & Linscott, reprinted 1954) recognized this kind of deeply organic worldview of generic interconnections as understanding of systemic complementarities underlying the development of complete scientific theory. In his words (op cit, p. 473):

> Scientific thought is a development of pre-scientific thought. As the concept of space was already fundamental in the latter, we must begin with the concept of space in pre-scientific thought. There are two ways of regarding concepts, both of which are necessary to understanding. The first is that of logical analysis. It answers the question, How do concepts and judgments depend on each other? In answering it we are on comparatively safe ground. It is the security by which we are so much impressed in mathematics. But this security is purchased at the price of emptiness of content. Concepts can only acquire content when they are connected, however indirectly, with sensible experience. But no logical investigation can reveal this connection; it can only be experienced. And yet it is this connection that determines the cognitive value of systems of concepts.

In our book thus far we understood the role of primitive concepts (Epistemology), their logical explanation (Ontology[1]) and experiences of system-entities (Ontic) to be cogently integrated in the complete derivation of a unique and universal methodology of scientific theory. This is our phenomenological model of unity of knowledge in the context of a neurocybernetic and system worldview. Such a generic interrelationship conveys the idea of complementary participation between systems and their entities in paired existences. We will refer to such embryonic pairing of participation in all things as complementarities. From it we derive the principle of pervasive complementarities in the phenomenology of unity of knowledge explaining the entire socio-scientific worldview and its methodology.

THE PRINCIPLE OF PERVASIVE COMPLEMENTARITIES AND
COOPERATION VERSUS COMPETITION

In the context of a neurocybernetic and system theory of scientific reasoning in everything, unity of knowledge is the foundational epistemology upon which the entire configuration and operation of a constructed world-system and its analysis rests. In the liberal worldview the opposed understanding of competition and cooperation as methodological concepts and their applications cannot attain the organic unity or pairing between systems and their entities. Relational domains cease to exist in the state of steady-state equilibrium or a time-dependent movement of such equilibriums. Pairing by learning is futile in the state of optimal allocation of resources. Enlargement of the resource allocation effect is realized by exogenous shocks. Such shocks as resource augmentation can continue the structural repetition of the system, but at a high cost. Endogenous relations defining learning system behavior cease to exist in any such terminal states (Shackle, op cit) even along the path of exogenous change that simply repeat structure. Learning instead is truly of the endogenous category in human behavior, choices, and the moral values of freedom, liberty and happiness. These are learned in paired (participatory) universes induced by unity of knowledge as the episteme.

The Liberal Mind of Individualism qua Competition versus Participation

In liberal economic and social theory the premise of social meaning for all these valued goals devolves on the individual as the ultimate unit of analysis. No further reduction is possible but simply the individual as the ultimate unit of analysis in any issue and problem under investigation. Such reductionism is the crisis of economic theory pointed out by Heilbroner and Milberg (1995). It results in contradiction between tenets and application of liberalism in economic methodology.

Such a problem of contradiction was pointed out by the great liberal thinker Frank Knight (1982 reprint, p. 60): "The essential social-ethical principle of liberalism or liberal individualism may now be stated, for the purpose of examination. It is that all relations between men ought ideally to rest on mutual free consent, and not on coercion, either on the part of other individuals or on the part of 'society' as politically organized in the state." The idea of absence of coercion of all types by the ascent of individualism is applied equally to the definition of morality, ethics, values and law, as to individual choices. In it the non-reducible nature of the divine laws in constructing a unique and universal worldview of unity of knowledge is rejected. Furthermore, Knight points out the single most contradictory aspect of liberal individualism. This is that "liberalism takes the individual as given, and views the social problem as one of right relations between given individuals. Individuals and their preferences of individualism become datum, whereas society, moral and ethical values are organic processes".

Such social processes can also lead into mutations, as in the case of social Darwinism, leaving behind conglomerates of competing groups and commodity bundles. The resulting groups and bundles subsequent to their mutations behave like competing individuals do. Institutions and enterprises are examples of such economic entities. They arise in the midst of knowledge-sharing. But subsequently, with the coming of creative destruction in the process of growth, efficiency and productive transformations, mutations into neoclassical type firms come about. Today, such competitions mark the growing specialization of a modernizing social and economic order in global capitalism (Peterson, 2003).

What is true of competition versus cooperation in liberal individualism, both as behavior and as system methodology, is extended to the realm of scientific methodology as well. Economic science is one of such theorizing and application of liberal individualism. In it the pursuit of happiness, liberty and freedom are explained as logical consequences of optimizing behavior and steady-state equilibriums. These states of the economic models of analysis are driven exogenously over time, or are endogenously regenerated by self-same preferences and behavior of agents and institutions. Nothing other than such a calculus of pleasure and pains exists in the hedonic worldview of economic liberalism. It gains its scientific refinements from the deepening of that methodology that differentiates economics as science from social reality, even in the name of modernist specialization. Weaver (op cit, p. 103) points out this one-dimensional nature of the liberal mind – "simplify man through various forms of scientific abstraction and to insist that he is 'nothing but' a thing that these techniques of exposition can explain."

In the context of a neurocybernetic and system theory of unity of knowledge, neither individualism nor mutations prevailing in the midst of competition can explain logically pairing behavior. Instead, there is simply continuous learning that spans over continuums of concepts, space and knowledge-induced entities. Consequently, the methodologies that are central to the logical explanation of steady-state and optimal equilibrium, either generated by liberal methodological individualism or mutations, cannot abide for the otherwise simulated evolutionary equilibriums in knowledge-induced fields of organic participation. Only complexity with unity of knowledge abides. Linear reasoning artificially created by simplicity ceases to be a methodological tool. The complexity of organic learning fields generates, and in turn, is sustained by the permanently and pervasively pairing relations in complementary world-systems.

Competition and adverse social behavior having a methodology contrary to the principle of pervasive complementarities can be explained by the same methodology of phenomenology of unity of knowledge. They now appear in the mathematical complementation (opposite) mode of analysis with different experiences within the phenomenology of unity of knowledge. The possibility of answering both the differentiated and the unitary embedded systems by the same methodology reflected in the project of phenomenology of unity of knowledge establishes this as a unique and universal paradigm of socio-scientific reasoning.

Contrarily, there is no concept of process in mainstream economic theory that could address unity of knowledge as a socio-scientific universality with process methodology in it. A process methodology in economic theory has been promoted as a much needed yet absent part of economic theorizing. But the methodology pointed out is still within the social Darwinian framework of mutations and conjectural evolution, which are known to be part and parcel of the neo-liberal paradigm. Nelson & Winter (1982) treat evolutionary economics within neoclassical orthodoxy. Georgescu-Roegen (1971) studies economic process as Popperian dynamics (Popper, 1998).

The Principle of Pervasive Complementarities in Participatory Systems

The principle of pervasive complementarities is a distinct axiom that is borne out by the logic of paired universes intra- and inter-systems by virtue of learning between these system-entities and to establish and sustain two fundamental behaviors. Firstly, there is the system behavior of interrelations between entities leading to unity of knowledge between them. Secondly, there is the behavior of maintaining the nature of stable systems by sustaining complementarities within these systems and by learning to adapt to such laws, rules, and guidance that reduce entropy. The principle of pervasive complementarities thus stands for relational epistemology that determines the learning processes of systems and their entities in all issues and problems of socio-scientific reasoning, methodological formalism and applications. Participation is signified by pervasive complementarities.

Competition is signified by cessation of learning and innovations in scientifically engineered equilibrium-optimal states. These happen under the assumptions of resource scarcity, quest for full information to define transitive logic of preferences as datum, and the exclusiveness of issues and problems within differentiated systems of analysis. The principle of pervasive participation across continuums and time is the strongest manifestation of unifying universes and their system-entities within the fold of unity of knowledge as the episteme.

IMPLICATIONS ON MORAL AND ETHICAL VALUES IN SYSTEMS DETERMINED BY CONTRARY SOCIO-SCIENTIFIC REASONINGS

Learning on the basis of the phenomenological worldview of unity of knowledge is seen to be the only methodology that departs from all theories and perceptions of neo-liberal vintage. In fact, we note two irrevocable conclusions regarding systems behavior arising from the generalization of a process-oriented and relational worldview in all scientific doctrines.

Firstly, scientific theory and logical analysis are disengaged from the organic reality, despite their methodical strength. This is the characteristic of socio-scientific reasoning in liberalism. It is extended to configure society and institutions, as also the natural sciences. Consequently, under such a liberal worldview, normative law, natural law and positive law are all governed by the same mind space of rationalism. In this framework of understanding the social meaning, if it so happens that human nature, and thereby social behavior, cannot be simplified as required by the methods of competition and methodological individualism, then as (Weaver, op cit, p. 103) writes, "we might not be able to manipulate him, and this thought is anathema to the positivistic party". On the same point of methodological individualism as the character of differentiated systems, Buchanan (reprinted 1999, p. 391) writes: "... it is evident that methodological individualism as a presupposition of inquiry, characterizes almost all research programs in economics and political science; constitutional economics does not depart from its more inclusive disciplinary bases in this respect."

The second inference we draw from our above discussions is that partial and limited interaction followed thereafter by mutations devolve into the same kind of methodological individualism. Methodological individualism and competition are now taken up between groups and institutions. Consequently, as Buchanan points out, there is no escape from the postulate of methodological individualism in all of socio-scientific reasoning under the liberal paradigm.

At many places in liberal socio-scientific reasoning, when the idea of cooperation and mutual interaction is used, this is the case in the framework of a methodology that is required as scientific nicety. It is needed to answer logical issues of behavior rather than real experiences and how such experiences can be formalized in terms of given laws and social guidance. Only the human will and rationalistic perceptions of social and scientific reality remain the ultimate explanatory factors.

Such configurations of the world-systems, human behavior and scientific theories in both the natural and social sciences being abstractions for logical handling of analytical systems, bring about particular perceptions on the meaning and role of morality, ethics and values. Buchanan sees the treatment of such themes as being outside the extant of the liberal worldview. Buchanan (op cit, p. 391) writes (note the edited word 'alone' at the end): "Individual evaluations are superseded by those emergent from God, natural law, right reason, or the state. This more subtle stance rejects methodological individualism, not on the claim that individuation is impossible, or that individual evaluations may not differ within a community, but rather on the claim that it is normatively improper to derive collective action from individual evaluations (alone)".

Therefore, the liberal philosophy and the emergent economic and social liberalism, the transmission of differentiated Darwinian mutation logic into scientific theory, run the serious problem of failing to comprehend the moral and ethical holism in the material world-system. They leave this department of human totality to exogenous impulses, which they assign to other differentiated domains of human inquiry. A system view of neurocybernetic is thereby rendered impossible in the differentiation process. To obtain a robust neurocybernetic and systemic understanding of morality, ethics, values, laws and ameliorative institutions of guidance, the project of scientific reconstruction must turn elsewhere.

Morality, ethics, values, laws and guidance being endogenous forces that can be continuously regenerated over continuums of knowledge-induced systems and their relational entities, fall in the project of unity of knowledge. In the sense of endogenous values, the material artifacts get embedded by such values. The inseparable relationship between values and material artifacts in the realm of commodities and preferences provides the conception of social and spiritual capital. The concept of value in this complex of interrelationship and its role in developing the understanding of ethical market exchange is essential in order to expand such valuations to ethical markets and ethicizing of markets. It is the process of ethicizing that establishes the learning process in a neurocybernetic and system framework.

The following elements of the learning process in unity of knowledge play the critical roles in the definition and measurement of value in market exchange: Knowledge-flows are derived and assigned ordinal values on the premise of unity of knowledge. The assignment problem of knowledge-flows within a combined natural system and institutionally controlled environment will be discussed in a later chapter. The laws and rules of moral and ethical guidance establish the human preferences and behavior in an ethicizing system. The discursive medium acts as the participatory venue in its extended sense of human agency in concert with the relational view of the natural system. Through such agential presence the rules and guidance are derived, the understanding of the ontological formalism is established, and the ontic measurements opened up both at the level of assigning ordinal values to knowledge-flows and at the level of inducing the state and policy (control) variables in the system. In this way, the entire system in which state variables appear in response to the implementation of preferences and behavior, as guided by the moral and ethical laws, rules and guidance, becomes sensitized by the epistemology, ontology and ontic framework of the IIE-process methodology.

VALUATION IN ETHICAL COMPLEXITY WITH PARTICIPATION CONTRA LINEAR SYSTEMS REASONING OF DIFFERENTIATED SYSTEMS

What is the theory of value from the endogenous moral and ethical standpoint in the complex system of learning in unity of knowledge? An example will lay down the ground work for formalization of the concept of Value.

A farmer returning from his work in the evening finds a rare fruit borne in a tree. He discovers it and finds after some research that the fruit has medicinal value. It is rare, having high medicinal value, and can therefore be sold to the affluent medical industry at an inordinate price. What price should the farmer charge in the two cases – firstly, in a competitive market system, and secondly, in an ethicizing market system, as explained above? What is the concept of Value linked to the commodity in these two systems?

Linear Reasoning in Competition Systems

In competitive market exchange the discovery and search of the medicinal fruit enters a monopoly for the farmer. Its buyer is the medical industry but its users are the sick. The industry itself can also charge a monopoly price for the medicine produced by the fruit. A monopoly price and trade-mark provision legitimates the pricing and control by the monopoly, despite that the monopoly can be partially regulated. Only those who can afford the monopoly price can access the medicine so generated by use of the rare fruit.

109

The farmer may raise more of such fruit and sell them to the medical sector while maintaining his monopoly power and causing monopoly power in the medical sector to persist. The needy sick are still not well off by the discovery and its continued production within the monopoly. Competition is the name of the game in certifying the practice of the monopoly pricing and profits. The value of the medicinal fruit is measured in terms of such a monopoly price and monopoly profits raised by the price, the output being restricted in the monopoly situation.

Complexity and the Theory of Value

In the participatory paradigm of unity of knowledge the ethical edicts of the learning system points to the discursive mechanism of natural system (fruit and its intrinsic value) and agential institutions (farmer, industry and polity) in understanding what ought to be the real value of the medicinal fruit. Note that the intrinsic value of the fruit has always been in it. It is the act of God. Therefore, this value cannot be fully claimed either by the farmer or the medical sector. Only the cost of labor, discovery, processing and transaction costs relating to the production and sale of the commodity can be included in the average cost of the fruit. Average cost pricing methods then result in the measurements of monopoly price, revenue and profits.

In the ethicizing market system the combination of systemic forces existing in the state of nature for the commodity with the guidance and preference formation of the conscious buyer and seller determines what ought to be the ultimate price of the commodity. Firstly, such an ethical price is determined by a discoursed 'fair return' to the seller and the industry. Subsequently, the expansion of production of the social good eliminates monopoly by an expanded participatory system of productions of the medicinal fruit and its medicinal products. Consequently, with production expansion, production diversification,

110

risk diversification and market penetration, the true value of the medicinal fruit in an ethicizing market exchange venue is approximated by the social wellbeing provided by the social good with its price, quantities, average cost, and institutional guidance by appropriate policies. All these together generate the knowledge-flows that qualify the preferences of participatory agents in an IIE-process of discourse involving market forces and ethical values.

A Simple Formalization of the Concept of Value

1. Competition System

Price is considered as the value of a good determined by the free play of market exchange. Also in the sense of consumer utility with substitute goods, price of a good is reflected as the marginal utility of a particular good among the substitutes. Consequently, in a perfectly competitive system price equals the marginal cost of production, and this reflects value. In the less than perfectly competitive markets price is determined along the average revenue curve corresponding to marginal revenue being equal to marginal cost determining profit-maximizing output. Value now is determined by the profit earned. Clearly now, as price increases, marginal utility declines but profits increase. Thus in a monopoly, the social welfare of the consumer is poised between these trade-offs between price and profit.

2. Participatory (Cooperative) System

Price is determined by production diversification, risk diversification and learning between market system and institutional laws, rules and social guidance. Consequently, price stabilizes and output increases. The wellbeing function of the consumer is now determined by price stabilization, output expansion, risk diversification and preference formation. These conditions are induced by discursive circular causation interrelationships between ethical learning (endogenous control variable) and market forces (endogenous state variables). Now the total change in social wellbeing by evolution of knowledge-flows reflects the Value concept and its measurement. These are determined in the ontic phase of quantitative simulation of the wellbeing function with circular causation relations between complementary system-entities. The wellbeing function depends upon co-evolved relationships between its complementary variables in the light of unity of knowledge. The latter is realized in learning experience by pervasive complementarities between these variables. In the evolutionary sense of learning the complementarities are generated and guided through the circular causation between system-entities in the simulation exercise.

NEUROCYBERNETIC AND SYSTEM PERSPECTIVE OF THE GENERALIZED MEANING OF VALUE

Beyond the sheer domain of ethical markets and economics are systemic interrelationships between such ethicizing systems and the entire socio-scientific order. Human comprehension and control of evolutionary learning systems is gained with increasingly levels of knowledge-flows premised on the epistemology of unity of knowledge and its ontological and ontic formalism. We must firstly understand what these evaluations of the individual and society mean in total social wellbeing. A deeper treatment of this critical concept and measurement of value in terms of knowledge-induced change of the total social wellbeing function is left for a later chapter.

In the example of the medicinal fruit and its valuation concept by means of the principle of pervasive complementarities in paired systems, we note that the unifying interrelations expand outwards and inwards, that is between systems and their entities. Neurocybernetic and system theory in this work always means the complex study of unifying circular causation relations in and between such two-dimensional domains with their multitudes of sub-systems and entities. In other words, neurocybernetic and system theory is inherently participatory in nature, hence pervasively complementary across space and time in the light of the episteme of unity of knowledge.

The intrinsic value of the medicinal fruit is premised on divine knowledge, a law primordially embedded in the fruit. From this primal knowledge embodied in an artifact the various systems learn how to use the precept of the divine laws to organize the wellbeing of society at large. The systems that learn are many. Among these we note the natural order (agriculture, markets, economy, society and productions), agents (farmer, industry, and polity), preferences and menus of the consumers and producers, respectively, and the IIE-process relations among all these, regenerated by means of ordinal valuations of learning along the evolutionary paths. The social wellbeing is the ontic formulation of the meaning of unity between these systems and their variables (entities). The interrelations between the complex sets of variables, the preferences and menus and the ordinal assignments of knowledge-values by interaction among all the represented systems, are explained by recursive relations between the entities. The social wellbeing function is thus simulated by means of the principle of pervasive complementarities between the variables of the social wellbeing function in the framework of unity of knowledge and by the method of circular causation relations between the system-entities.

Neurocybernetic and system theory cannot exist in the case of linear reasoning, as of the neo-liberal orders and their reflection on its attenuating socio-scientific order. Linear reasoning is a particular degenerate derivation of the generalized model of phenomenology of unity of knowledge. The principle of pervasive complementarities does not exist in linear reasoning, because either competition between individual forms or mutated systems of entities pervade in social welfare function (as opposed to social wellbeing criterion).

Consequently, circular causation cannot exist because the learning and policy impacts remain exogenous between them and to market forces. So also are technology and many other-system influences exogenous to the rest. On the other hand, if circular causation does exist for a purely endogenous system of variables the result will be greater degree of mutations out of competing forms. Mutation conveys the meaning of complexity in such disequilibrium learning models. Unity of knowledge is impossible in the received socio-scientific theories in their entirety. We have discussed this point earlier.

CONCLUSION:
RECASTING HUMAN FUTURE IN COMPUTING REALITY

The contrast between complex and linear reasoning is brought out in this chapter in two distinct ways. Firstly, non-participatory and competing systems do not have the particular methodology, and therefore, cannot be used to explain complex relations in the context of learning by unity of knowledge. Secondly, even when complexity is defined, the complex systems result in Darwinian type mutations. Mutations subsequently become competing forms in groups and alliances. Mutations are thereby either of the nature of fragmented competing forms or exist in a perpetual order of entropy by competition, conflicts and bifurcations. The episteme of unity of knowledge is thereby denied.

The advent of neo-liberalism and the socio-scientific systems it influences leaves the human future in a replete of the past. Modernity and post-modernity now become differentiated pairs of anomie within themselves. Capitalism is a stark example of such anomie respecting social issues and the social economy on human future. Global capitalism is fired by competing or mutating-competing or simply anomie relations between groups of nation states.

The broadest division comprises the industrialized and developing world but with many world-systems inside these as well. Globalization is the result of global networking of economic, social and culture-ideology systems that relent on what the underlying episteme of these various systems dictate. For capitalist globalization, the ever presence of competitive forces in the description of networking between economic, social and culture-ideology forces is always of the types mentioned above. Yet globalization as a human future can also be quite contrary. That is the message of the epistemological, ontological and ontic reformulation of systems thinking. In this regard, Sklair (2002, p. 84) writes: "Globalizing processes are abstract concepts, but the transnational practices that create them refer directly to what agents and agencies do and derive meaning from the institutional settings in which they occur, and because of which they have determinate effects."

Human sustainability in the complete sense of systemic sustainability and institutional and value sustainability cannot be attained in the fragmented framework of capitalist globalization due to the preconditioning of such disenfranchised world-systems and their neo-liberal abstractions in capitalist globalization. Therefore, the human future of sustainability in thought, behavior and organization of life must be sought for elsewhere. Neurocybernetic and system theory is premised on the methodological worldview of unity of knowledge . This defines the phenomenology that beckons the alternative of human future by virtue of the inherently sustainable nature of its dynamics. We will discuss many of these critical issues in the subsequent chapters. They were also covered in the previous ones.

In the end, the project of *Computing Reality* now points to the building blocks of a neurocybernetic and system theory of unity of knowledge that can make participatory system-forms of learning as the basis of human sustainability. The principle of pervasive complementarities premises a new organizational conception interrelating markets (state variables and socio-scientific systems) and institutions (control variables and polity as organizational systems). In such relations knowledge-induced preferences and behavior pervade in the process of evolutionary learning. Thus the methodological entirety and the human worlds are both uniquely and universally constructed by the same worldview of IIE-process methodology premised in the phenomenology of unity of knowledge.

NOTES

[1] Gruber's definition of Ontology is the one most helpful for the understanding of phenomenology in neurocybernetic and system theory of learning systems premised in unity of knowledge, which is the project of *Computing Reality*. He defines Ontology in the following words (Gruber, 1993, internet version):

> Ontology is an explicit specification of a conceptualization. .... For AI systems, what "exists" is that which can be represented. When the knowledge of a domain is represented in a declarative formalism, the set of objects that can be represented is called the universe of discourse. This set of objects, and the describable relationships among them, are reflected in the representational vocabulary with which a knowledge-based program represents knowledge. Thus, in the context of AI, we can describe the ontology of a program by defining a set of representational terms. In such ontology, definitions associate the names of entities in the universe of discourse (e.g., classes, relations, functions, or other objects) with human-readable text describing what the names mean, and formal axioms that constrain the interpretation and well-formed use of these terms. Formally, ontology is the statement of a logical theory. (edited)

T.R. Gruber. 1993. "Toward principles for the design of ontologies used for knowledge sharing", In N.Guarino and R.Poli, editors, Formal Ontology in Conceptual Analysis and Knowledge Representation, Padova, Italy, International Workshop on Formal Ontology, Kluwer Academic Publishers.

REFERENCES

Bohr, N. 1985. "Discussions with Einstein on epistemological issues", in H. Folse, *The Philosophy of Niels Bohr: The Framework of Complementarity*, Amsterdam, The Netherlands: North Holland Physics Publishing.

Buchanan, J. M. 1999 (reprint). "The domain of constitutional economics", in *The Collected Works of James M. Buchanan, Vol. 1: The Logical Foundations of Constitutional Liberty*, Indianapolis, IN: Liberty Fund, pp. 377-395.

Choudhury, M.A. 1996. "Markets as a System of Social Contracts", *The International Journal of Social Economics*, Vol. 23, No.1.

Einstein, A. 1954 reprinted. "The problem of space, ether, and the field in physics", in Commins, S. & Linscott, R.N. eds. *Man & the Universe: The Philosophers of Science*, New York, NY: The Pocket Library, pp. 473-484.

Einstein, A. undated. "The laws of science and the laws of ethics", in his *Essays in Physics*, New York, NY: Philosophical Library.

Fukuyama, F. 1992. *The End of History and the Last Man*, New York, N.Y: The Free Press

Fitzpatrick, T. 2003. "Postmodernism and new directions" in Alcock, P. Erskine, A. and May, M. eds. *The Student's Companion to Social Policy*, Oxford, UK: Blackwell Publishing, pp. 127-136.

Georgescu-Roegen, N. 1971. *The Entropy Law and the Economic Process*, Cambridge, MA: Harvard University Press.

Gruber. T.R. 1993. "Toward principles for the design of ontologies used for knowledge sharing", In N. Guarino and R. Poli, editors, *Formal Ontology in Conceptual Analysis and Knowledge Representation*, Padova, Italy, International Workshop on Formal Ontology, Kluwer Academic Publishers.

Hammond, P.J. 1987. "On reconciling Arrow's theory of social choice with Harsanyi's fundamental utilitarianism", in Feiwel, G. ed. *Arrow and the Foundations of the Theory of Economic Policy*, London, Eng: Macmillan, pp. 179-222.

Heidegger, M. trans. Hofstadter, A. 1988. "The thesis of modern ontology: the basic ways of being are the being of nature (res extensa) and the being of mind (res cogitans)", in *The Basic Problems of Phenomenology*, Bloomington & Indianapolis, IN: Indiana University Press, pp. 122-224.

Heilbroner, R. & Milberg, W. 1995. *The Crisis of Vision in Modern Economic Thought*, Cambridge, UK: Cambridge University Press.

Intrilligator, M.D. 1971. "The control problem", in *Mathematical Optimization and Economic Theory*, Englewood Cliffs, NJ: Prentice-Hall, Inc. pp. 292-305.

Knight, F. H. 1982 (reprint). *Freedom & Reform, Essays in Economics and Social Policy*, Indianapolis, IN: The Liberty Fund.

Kuhn, T. 1970 reprinted. "The structure of scientific revolution", in Neurath, O, Carnap, R. & Morris, C. eds. *Foundations of the Unity of Science, Towards an International Encyclopedia of Unified Science, Vol. II*, Nos. 1-9, Chicago, ILL: University of Chicago Press, pp. 55-236.

Lancaster, K.J. 1966. "A New Approach to Consumer Theory". *Journal of Political Economy*, Vol. 74, pp. 132-157.

Lyotard, J-F, 1984. *The Postmodern Condition*, Manchester, UK: Manchester University Press.

Minogue, K. 1963. *The Liberal Mind*, Indianapolis, IN: The Liberty Fund.

Nelson, R.R. and S.G. Winter, 1982. *An Evolutionary Theory of Economic Change*, Cambridge, MA: The Belknap Press of the Harvard University Press.

Petersen, V.S. 2003. "Productivity", in *A Critical Rewriting of Global Political Economy*, London, Eng: Routledge, pp. 44-76.

Palan, R. 2000. "New trends in global political economy", in Palan, R. ed. *Global Political Economy*, London, Eng: Routledge, pp. 1-18.

Parkin, M. & Bade, R. 1991. "Talk with Assar Lindbeck", in *Economics, Canada in the Global Environment*, Don Mills, Ont: Addison-Wesley Publishers, pp. 1-4

Popper, K. 1998. *Conjectures and Refutations: The Growth of Scientific Knowledge*, London, Eng: Routledge & Kegan Paul.

Sen, A. 1999. *Commodities and Capabilities*, Amsterdam: Elsevier.

Shackle, G.L.S. 1972. *Epistemics and Economics*, Cambridge, Eng: Cambridge University Press.

Sklair, L. 2003. "Transnational practices: corporations, class, and consumerism", in *Globalization, Capitalism and Its Alternatives*, Oxford, Eng: Oxford University Press, pp. 84-117.

Smith, A. [1753] 1966. The *Theory of Moral Sentiments*, New York, NY: Augustus M. Kelley.

Smith, A. [1776] 1937. The *Wealth of Nations*, Cannan, E. ed. Book. 1. Chicago, ILL: University of Chicago Press.

Vanek, J. 1971. "The participatory economy in a theory of social evolution", in his *The Participatory Economy: an Evolutionary and a Strategy for Development*, pp.51-89, Ithaca, NY: Cornell University Press.

Wallerstein, I. 1998. "Spacetime as the basis of knowledge", in Borda, O.F. ed. *People's Participation, Challenges Ahead*, New York, NY: The Apex Press, pp. 43-62.

Wallerstein, I. 1974. *The Modern World Systems*, New York, NY: Academic Press.

Weaver, R.M. 1977. "Individuality and freedom", in Morley, F. ed. *Essays on Individuality*, Indianapolis, IN: The Liberty Fund, pp. 89-113.

TECHNICAL APPENDIX TO CHAPTER 4:

MULTIDIMENSIONAL SYSTEMS PHENOMENON

Neurocybernetic and system theory deals with multidimensional spaces of interaction, integration and evolutionary dynamics as in the IIE-process methodology of unity of knowledge. We display this scenario below in connection with market forces (state variables), and institutions (control variables, policy variables) under the impact of learning (knowledge-induced preference formation).

Matrix $[x_{ij}]$ =   (A4.1)

i	1	2	3	...	n		Vector Notation
market forces							
j							**Vector Notation**
Institutional forces							
1	x_{11}	x_{12}	x_{13}	...	x_{1n}	=	$\mathbf{x}_{1j}$
2	x_{21}	x_{22}	x_{23}	...	x_{2m}	=	$\mathbf{x}_{2j}$
3	x_{31}	x_{32}	x_{33}	...	x_{3n}	=	$\mathbf{x}_{3j}$
....							
m	x_{m1}	x_{m2}	x_{m3}	...	x_{mn}	=	$\mathbf{x}_{mj}$

Case 1: Economic and Social Competition

1. With Methodological Individualism

Preferences are disjoint and the utility function in market forces competed for goods and services and in agential institutions with competing individuals, the total utility function (U) is defined in terms of individuated utility functions as,

$$U = \Sigma_{i=1}^{n}\Sigma_{j=1}^{m} U_{ij} (x_{ij})[\wp(\theta = \phi)]$$ (A4.2)

2. Mutations in Group Specific Methodological Individualism

The same formulation as in (A4.2) holds except that the matrix [xij] is now converted into sub-matrixes [x_{kl}], $k \in \{i\}$; $l \in \{j\}$, in terms of the number of competing groups of individuals $\{k \in \{i\}\}$ and over competing commodities $\{1 \in \{j\}\}$. The symbols are accordingly redefined. Preferences with null learning, $\wp(\theta = \phi)$, as non-existing interaction by process in states of steady-state equilibrium and optimal allocation of resources holds.

Case 2: Participatory Systems, Learning between Goods and Individuals

1. Pervasively Complementary Preferences

$$\wp(\theta \neq \phi) = \cap_{ij}\{ \wp_{ij} (\theta \neq \phi)\}, \qquad\qquad\qquad (A4.3)$$

with θ = probability limit $_{\text{Interactions}} [\cap_{ij}\theta_{ij}] \neq \phi \qquad\qquad (A4.4)$

The wellbeing function is now defined contrary to the utilitarian form in expression (A4.2) as,

$$W = \cup_{\text{interaction}} \cap_{ij} W_{ij} (x_{ij})[\wp(\theta \neq \phi)] \qquad\qquad (A4.5)$$

$$dW/d\theta = d/d\theta(\cup_{\text{interaction}} \cap_{ij} W_{ij}/dx_{ij}).(dx_{ij}/d\theta) > 0 \qquad (A4.6)$$

as evolutionary property implying each of the agent-commodity indexed wellbeing, W_{ij}, and the system-entities (variables) co-evolve in the same direction of evolutionary learning.

In the set of equations (A4.3) – (A4.6) the probability limit of expression (A4.4) will also make the variables and functions to be probabilistic in nature, so that only knowledge-induced evolutionary forms can exist over sequences of IEF-processes.

The simulation of (A4.5) is carried out over (m+n)-number of circular causation relations between the {xij}-variables as system-entities, plus the epistemic primal, $\theta \in \Omega$, as the topology of unity of knowledge defined by stated laws and guidance on unity of knowledge, e.g. the derivation of the principle of pervasive complementarities. (A4.7)

Neurocybernetic system (NS) in unity of knowledge is defined by,

NS = {matrix [xij], given the conditions (A4.3) – (A4.7)} (A4.8)

Computing reality is explained by the phenomenology of unity of knowledge comprising the entire simulation problem included in expression

(A4.8). A thorough characterization of the above kinds of systems will be done in subsequent chapters.

CHAPTER 5: PHENOMENOLOGY AS CONSCIOUSNESS IN LEARNING SYSTEMS

Learning systems have artificial intelligence as well as social intelligence. In the neurocybernetic and system theoretic worldview of the interactive, integrative and evolutionary processes of knowledge-flows the model of phenomenology of unity of knowledge is organized in specific ways. These are different from the phenomenological ideas presented in the literature. Thus the project of phenomenology of knowledge must be first understood both in its classical conception and in the context of unity of knowledge.

The objective of this chapter is to review the classical literature on phenomenology of knowledge and then to show how this area differs substantially from the one on phenomenology of unity of knowledge. Consequently, we will examine how the neurocybernetic and system theory is distinctly formulated in these two different methodological contexts. The contents of this chapter belong to the field of philosophy of science, which will be invoked to show how a learning system theory can be developed for the case at point, that is, money and real economy linkage.

Furthermore, the philosophical and methodological contents of this chapter will open doors to the climactic chapter 6 on formal logic of learning systems in the light of the phenomenology of unity of knowledge. This chapter is thus the gateway to the formalism of the unique and universal worldview premised on divine oneness as it can be understood within its epistemological context differently from the rationalistic model of classical phenomenology. Our methodological and philosophical focus on the study of phenomenology of unity of knowledge is on the topic of consciousness. The equivalent question thus opened up is this: How does artificial intelligence integrate with the human world in terms of consciousness that spans the entire socio-scientific world-systems?

THE PROJECT OF HUMAN CONSCIOUSNESS AND PHENOMENOLOGY

Consciousness and Phenomenology

Encyclopedia Britannica (1981) defines the project of phenomenology in terms of human consciousness as given below. We use here Husserl's approach to explain the reduction of reality to self-perception, leaving out for a separate section the meaning given to phenomenology and consciousness by Martin Heidegger.

Husserl's conception of consciousness and phenomenology is a search for the most reduced premise that can explain the totality of human experience. The project aims at isolating such interconnection between objects and the self to the sole origin of reason in the beholder and not to any external episteme. From this kind of relationship a self-perception, fully liberated from preconditions and axiomatic references, is developed to construct the relations and conceptions of the relationship of self with the other. Such is the ontological stage of constructing relational understanding of a perceived idea regarding the subject that is intentionally objectified by the human perceiver. Encyclopedia Britannica (op cit, p. 212) writes,

> The phenomenological investigator must examine the different forms of intentionality in a reflective attitude because it is precisely in and through the corresponding intentionality that each domain of objects becomes accessible to him. Husserl took as his point of departure mathematical entities and later examined logical structures, in order finally to achieve the insight that each being must be grasped in its correlation to consciousness, because each datum becomes accessible to a person only insofar as it has meaning for him. From this position, regional ontologies, or realms of being, develop .... Moreover, Husserl distinguished formal ontologies – such as the region of the logical – from material ontologies.

126

Through the combination of self-conceived perceptions of reality and the material ontological regions both in terms of experiences and relations, the conception of phenomenology is defined (Encyclopedia, op cit, p. 210): "as a philosophical movement the primary objective of which is the direct investigation and description of phenomena as consciously experienced, without theories about their causal explanation and as free as possible from unexamined preconceptions and presuppositions."

We note that the above scope of the phenomenology project and its link with human consciousness is somewhat different from the way we understand consciousness and phenomenology in this work. Contrary to the experiential and object oriented intentionality of the ego as perceiver our understanding of phenomenology is to construct an object by epistemological reference followed by its corresponding ontological construction and ontic application and inferences. Thus phenomenology in our study comprises all of these three stages of total knowledge formation. This totality assigns meaning to specific issues and problems in the constructed world-system.

By the starting point of regional ontology pertaining to specific issues and problems of investigation Husserl's phenomenology rejects the premise of the transcendental law. Our meaning of phenomenology cannot accept this communal picture of self-perception, as the search then evades a logical reduction of all systems of intellection and knowing to the most irreducible premise, except in self as the ultimate ego. Contrarily, we argue that the most irreducible limit of possibility is the premise of the divine laws. It also marks the possibility that results in the determination of the unique and universal worldview of unity of knowledge. This is the nature of the phenomenology of unity of knowledge taken up in the neurocybernetic and systems theoretic framework.

Consciousness in our work is determined by relational unity of knowledge

127

and evolutionary circular causation in the process of learning based on this episteme of unity of knowledge. Consciousness in Husserl's conception is determined by the rationalistic position of intentionality within self in relation to the world. Such consciousness can lead to myriads of different ontologies for the same issues and problems under investigation, or the unique regime of the ontology must be governed and enforced for achieving consensus by the will of a dominant player. An example here is of Rawlsian minimax game that Wolff (1977) points out can be terminated only by the presence of an ethical preceptor. The supreme governance and enforcement are also exhibited in the case of existence of a dictator to well define Arrow's social welfare function (Arrow, 1951).

The theme of consciousness and how it leads into socio-scientific formalism is an inquiry on rationalism in classical terms and on the divine origin understood as the universal One in our project. This means that consciousness as human origin, nature and capacity for knowledge is reduced in its scope of comprehension, if it cannot encompass the formal nature of divine oneness, bringing this episteme to the order of man and artifacts within the learning world-systems only. Only by overarching between the domains of abstraction and evidential forms it is possible to universalize the scope of consciousness and to enable it to connect between self and others, both in terms of the human world-system and that of its artifacts (computing as example of machine language) (Choudhury, 1998).

IMMANUEL KANT ON CONSCIOUSNESS

Contrary to this universalistic nature of consciousness according to the episteme of the divine laws is the idea of consciousness according to Rationalism. Rationalism can be explained in terms of Carnap's (1966) schema on the problematique of Kant's problem of heteronomy. Kant wrote on Pure Reason as the domain of human consciousness, wherein the divine laws and the divine essence of reality reside. In this sense, Kant was a moralist, a personality that is projected in his words (Kant, 1949, p. 261):

> Two things fill the mind with ever new and increasing awe and admiration the more frequently and continuously reflection is occupied with them; the starred heaven above me and the moral law within me. I ought not to seek either outside my field of vision, as though they were either shrouded in obscurity or were visionary. I see them confronting me and link them immediately with the consciousness of my existence.

Yet as Kant wrote on the limitation of Practical Reason he dichotomized it from Pure Reason. Kant treated all that was deductively derived from the origin of Pure Reason *a priori* to be essential knowledge. All that was deduced inductively from the origin of Practical Reason *a posteriori* was not essential knowledge. In this regard Kant wrote,

> This, then, is a question which at least calls for closer examination, and does not permit any off-hand answer; whether there is any knowledge that is thus independent of experience and even of all impressions of the senses. Such knowledge is entitled a priori, and is distinguished from the empirical, which has its sources a posteriori, that is, in experience.
> ....
> In what follows, therefore, we shall understand by a priori knowledge, no knowledge independent of this or that experience, but knowledge absolutely independent of all experience. Opposed to it is empirical knowledge, which is knowledge possible only a posteriori, that is through experience.

Kant is thus pointing out the duality between Pure Reason and Practical Reason; that is between the *a priori* and the *a posteriori*, between mind and matter, between the deductive and inductive, and the normative and positive laws. It is non-substantive for knowledge formation to reverse this relational order, for experience is denigrated as not being the seat of deductive and *a priori* reasoning. The evolutionary relational epistemology between pure reason and practical reason is thus denied due to the absence of circular causation between these disjoint Kantian realms. Such a duality causes the problem of heteronomy despite the critical imperative of the *a priori* that is embedded solely in pure reason. Consequently, since God resides in Kant's Pure Reason and is *a priori* to everything of experience in the *a posteriori* of Practical Reason, therefore the correspondence between abstraction and materiality is seen to commence from the domain of *a priori* to the domain of *a posteriori* and not vice versa.

This is not all. As Carnap (op cit) pointed out, the region of Pure Reason is unable to comprehend the divine laws due to human intervention as the *a priori*. Consequently, the correspondence from Pure Reason within the realm of the divine laws cannot be mapped on to the region of Practical Reason, the *a posteriori* experience. The problem of heteronomy is this discontinuity between the *a priori* of the divine laws as the most supreme categorical imperative and the experiential world of the *a posteriori*. As well as, it is the problem of impossibility for correspondence to exist between the *a priori* par excellence of divine niche and the rest of the region of Pure Reason. Upon this is also compounded the problematique of absence of circular causation (feedback relations) between the *a posteriori* domain and the *a priori* domain. On the problem of impossibility of mapping the divine laws in its pure form except by a reason caused by human intentional perception, Kant (1963, pp. 80-81) wrote,

A clear exposition of morality of itself leads to the belief in God. Belief in this philosophic connexion means not trust in a revelation, but trust arising from the use of the reason, which springs from the principle of practical morality.

Religion is the application of the moral laws to the knowledge of God, and not the origins of morals. For let us imagine a religion priori to all morality: then this would imply a relation to God, and would therefore consist in recognizing Him as a mighty lord whom we should have to placate. All religion assumes morality, and morality cannot, therefore, be derived from religion......

Thus Kant's dichotomy between *a priori* and *a posteriori* knowledge becomes a dysfunctional factor in realizing the full power of human consciousness. According to the methodology so constructed, only deductive reasoning remains primal in knowledge formation. It is absolutely impossible to cast the divine laws on to experience due to the problem of heteronomy that partitions the niche of God within the realm of Pure Reason from any relationship between Pure Reason and Practical Reason (Choudhury, 1997). The divine laws thus remain dysfunctional in Kantian consciousness. The methodology of categorical imperative thus devolves into Rationalism. The universality and uniqueness of the meaning of worldview (weltanschauung), extending beyond the sheer precincts of human ego to include the episteme of unity of knowledge, cannot be realized in Kantian phenomenology. The rationalistic conceptions of Kant are borne out in his writings (Kant, trans. Infeld, 1963; Kant, trans. Paton, 1948).

HUME'S INDUCTIVE ONTOLOGY AND CONSCIOUSNESS

Hume's ontological principle depended on inductive reasoning as the primal source of knowledge. Experience as in the *a posteriori* sense was considered as the premise of knowledge in reference to the sensate world. Hume (reprinted 1992, p. xvii) wrote on this account:

For me it seems evident, that the essence of the mind being equally unknown to us with that of external bodies, it must be equally impossible to form any notion of its powers and qualities otherwise than from careful and exact experiments, and the observation of those particular effects, which result from its different circumstances and situations. And tho' we must endeavour to render all our principles as universal as possible, by tracing up our experiments to the utmost, and explaining all effects from the simplest and fewer causes, 'tis still certain we cannot go beyond experience, and any hypothesis, that pretends to discover the ultimate original qualities of human nature, ought at first to be rejected as presumptuous and chimerical.

In Hume's ontological principle of experience, neither can we accommodate the origin of reason and construction of experiential world-systems to the divine laws, as in the case of the phenomenology of unity of knowledge, nor can we interconnect the deductive with the inductive. This was also the problem of discontinuity between the two premises of reasoning in Kant. Both Kant and Hume thus abandon the project of *functional* relevance of the divine laws in the experiential world-systems. Knowledge in all such dichotomous systems cannot arise from any unique and universal worldview due to the abundance of issues and problems that are left out from the social tenets that emanate from the consciousness based on the divine order. Man, science and social constructions are all thereby left to the vagaries of differentiated human egos and perceptions. On this matter wrote Bergson (in Russell, p. pp. 756-65), "Perception is the lowest state of mind. It is like mind without memory."

The problem of rationalistic dichotomy in reasoning is not the intentioned consequences of otherwise well-intentioned human minds. Rather the problem of dichotomy, discontinuity and heteronomy is caused by the methodological consequences of such dichotomous forms of reasoning. Dichotomous reasoning of such types, namely between deductive and inductive reasoning, between matter and mind, and between God and existence is the sole cause of the methodological impossibility for unity of knowledge in such thinking realms. Consequently, neither the unified human consciousness nor the phenomenology of unity of knowledge can be derived as a methodological consequence from such dichotomous thinking modes. This is despite the fact that the best of minds have continued to address the theme of unity of knowledge in their works.

The building of a neurocybernetic and system theory based on unity of knowledge across complex relations also gets blocked by the design of complex relational orders, which otherwise the phenomenology of unity of knowledge seeks to address. Neurocybernetic and system theory would deal with complex relations rather than linear systems as a rule, by virtue of the complex realism in the project of unity of knowledge. A complex relation cannot be derived by any degree of manipulation within a linear system. For instance, linearly adding two mathematical functions to result in another is to ignore the fact that there is a region of 'extension' (extant) between these functions that make them interact. Consequently, with this idea of extension of abstract ideas in view, the two functions must be expressed as product function that enable interaction and show up in the resulting functional form.

Another example is that of color mixing. Note the following impossibility in a linear mixing context: It is known that Red + Green = Yellow (as linear addition); Blue + Green = Cyan; Red + Blue = Magenta; Red + Green + Blue = White; Cyan + Magenta + Yellow + Blue = Gray. Yet, if we were to linearly add these combinations we note that whereas, Cyan + Magenta + Yellow + Blue = Gray, yet the common elements, Red + Green + Blue = White. Consequently in such a case, a linear combination of colors gives us two different results. Color is thus an 'extended' quality based on the properties of light and things. Such extensions overlap different colors. Consequently, even though say, Red and

Green combine to form Yellow, yet these two colors are within Yellow as composites.

Yet another example from mathematics is this. Imaginary numbers do not exist in the experiential order of computation. They are sheer abstractions resulting from the solutions of polynomial algebraic expressions. Thus it is meaningless to interpret, $(\sqrt{-1} - \sqrt{-1}) = 0$, as the subtraction of two non-existing entities. Yet it is mathematically useful to define the concept of 'zero' **(0)** by the expression, $f(\sqrt{-1}) = \mathbf{0}$. In this case, imaginary numbers must be converted to real numbers by rationalization, which now conveys experiential measures by the process of converting imaginary functions into real ones.

In this way, many such examples can be developed to show that computing reality is a real world experience that exists in 'extended' (extant) form. Thus only complex functions (compound transformations) exist. These complex relations extend across interrelating phenomena as compounded linear or compounded non-linear functions and relations.

On the theme of working with imaginary number, Wiener (1961, p. x) points out that, when two trigonometric functions with non-linear coefficients, as in the case of learning coefficients, are multiplied with each other, the properties of linear groups of functions are altered. For instance, the compounding or multiplication with the family of functions $(A(\theta,t).\cos(\theta_t) + B(\theta,t).\sin(\theta_t))$, for different values of the learning parameter $\{\theta\}$ over time 't'. 't' is treated here as the recorder of events that in our case are induced by knowledge-flows premised on the episteme of unity of knowledge.

The resulting multiplicative family becomes a non-linear family of trigonometric functions. Thus the nature of learning coefficients $(A(\theta,t), B(\theta,t))$ makes the non-linear difference in the family of multiplicative functions here. If this property of extension is not invoked, so that the coefficients do not share in

134

the common θ-value over time, or if there was just the time variable 't', without θ being common to the trigonometric values, then linearity will persist. These points are raised by Wiener in his cybernetic formalism.

Yet the reverse question needs to be addressed: Is it possible to deduce linear forms from complex ones? This kind of disaggregation is not possible in the strict sense of extended fields of relations between functions. It is possible only in the local sense of reducing such extended relations to zero and to cause an end to systemic novelty otherwise caused by learning. A profound example in the social world is that of methodological individualism or hegemonic governance. In the natural sciences, the example is of reducing initially given biological interaction into mutated forms. The latter is the case of social and anthropological Darwinism (Dawkins, 1976).

Yet Hume would have liked us to believe contrarily. On the matter of disaggregation of complex ideas (non-linearity) into simple ideas (linear Relations, Modes and Substances) Hume writes (op cit, p. 13):

> Amongst the effects of this union or association of ideas, there are none more remarkable, than those complex ideas, which are the common subjects of our thoughts and reasoning, and generally arise from some principle of union among our simple ideas. These complex ideas may be divided into Relations, Modes and Substances.

On the matter of aggregation of simple ideas (linearity) into complex ones (non-linearity) Hume writes (op cit p. 16):

> The idea of a substance as well as that of a mode, is nothing but a collection of simple ideas, that are united by the imagination, and have a particular name assigned to them, by which we are able to recall, either to ourselves or others, that collection.

EDMUND HUSSERL CONTRA KANTIAN METAPHYSICS

Edmund Husserl (1964 trans. Alston & Nakhnikian) wanted to reform the dichotomy between deductive and inductive reasoning, between the gap of Pure Reason (*a priori* premise) and Practical Reason (*a posteriori* premise), as is the case with Kant's heteronomy and Hume's ontological principle. As a critique of Kant on the subject of the Kantian heteronomy between noumena (*a priori* premise) and phenomenon (*a posteriori* premise) Hammond et al. (1991, p. 87-126) write regarding the contrast (edited): "He (Husserl) notes two distinguishing features between the two types of transcendental idealism. First, Kant retains the idea of a transcendent realm: this is Kant's world of things in themselves (noumena). Belief in such a world, for Kant, is intelligible despite its being impossible to know such a world. Husserl rejects this idea of a transcendent realm. For Husserl the only meaningful idea of transcendence is to be eliminated from philosophy."

Husserl thereby resigns to the omission of the transcendence wherein the realm of God the divine exists. His phenomenology is thereby to limit the ontological principle to sense perception as the starting point of knowledge. The recursive learning of the IIE-process methodology is left wandering in uncharted paths of the human ego, rationalism and perceptions, without having an initial premise and a final destination to relent upon. The phenomenological world-system of self and the world commences from human rationalism, and thereby from a plethora of differentiated epistemes. None of these has a unique determination. Consequences of over-determination as we found in Marxist political economy, and the conflictual world-systems of liberalism premised on methodological individualism and driven by competing mindspaces, become the order of socio-scientific conceptions and constructions. Within such an ontological construction relenting on the primacy of the sensible world, as in

Hume as well, Husserl wants to combine the system by a recursive process of causality. Yet this remains contained in the entireties of the rationalist world-systems and mindspaces.

Kant's phenomenology rests on the ultimate primacy of the transcendental Ego. Kant accepts this ultimate realm to be God's. Yet Kant is unable to relate the meaning of God with experience. Hence Kant's Pure Reason despite accepting the ultimate phenomenology of unified consciousness to rest on God, leaves this fact as a metaphysical entity. God thus remains independent of experience, and is thereby dysfunctional in the development of the ontological principle. The matter of the origin and the possibility for knowledge is thereby left to the human ego, the Dasein of self and individuality. Husserl's phenomenology of unified consciousness resides in the object of experience independently of the transcendental realm. Matters of God, spirit and soul do not enter Husserl's phenomenology. Transcendental Ego of this type is circularly contained within the ontologically constructed human experience as the ultimate ego of cause and effect. To Kant, now independent of God as the primal cause of all causation once the epistemological premise is rightly understood, the world of objects is constructed ontologically and experientially according to those epistemological *a priori*, namely noumena projected on to phenomena. Throughout this book we have interpreted the meaning of ontology in the engineering sense as pointed out earlier, and not in the metaphysical sense because of its non-functional nature in formalizing a methodology.

Hammond et al (op cit, p. 93-94) write in this regard on Kant and Husserl's opposed views on the idea of categorical imperative derived from the premise of noumena:

> In this sense, for example, the beginning of the universe, an event which in principle it is important to have any experience of, would be a noumenal event. If we try to achieve knowledge of this event then we are

bound to fail: for knowledge is possible only if it is possible to relate concepts to experience. Concepts which do not relate to experience are transcendent concepts, e.g. God, or the soul, and they refer to noumena.

Husserl objects to this division between a world which can be known and a world which cannot be known. Like Kant, Husserl rejects the idea of what transcends all experience as a possible object of knowledge, because it is impossible to experience, but unlike Kant, he wants to eliminate such a concept from having any role to play in knowledge.

The cybernetic idea of extension in the realm of phenomena applies both to Kant and Husserl. Both consider the concept of noumena, the realm of the categorical imperative and the seat of Pure Reason to be within human experience. In both, the substantial questions of the divine laws, of God and the phenomenology of unity of knowledge formalized and enabled by the divine laws remain outside human inquiry. This is the great stumbling disability of all Occidental epistemologists and phenomenologists. Thereby, the same kind of philosophy of science has pervaded throughout Occidental socio-scientific thought and its practice in the scholarly disciplines and institutional orders.

The extension of the realm of noumena and phenomena within the domain of experience is analogous to our understanding of the total meaning of phenomenology. Yet such other theories cannot establish the phenomenology of unity of knowledge. Our groundwork of the project of phenomenology of unity of knowledge underlying a universal and unique socio-scientific worldview formalizes and applies the divine order within the experiential world-systems, which both Kant and Husserl failed to incorporate in their methodologies. Thereby, Kant and Husserl's theories leave unattended, unexplained and untapped a great portion of human thought and its 'concrescent' role between noumena and phenomena as undivided fields of learning resting upon the episteme of unity of knowledge. In a subsequent chapter we will examine particularly the formal model given by Sztompka (1991) on such a methodology, yet dispensed from the realism of the divine origin, and thereby becoming dysfunctional in the project of unity of knowledge as the unique and universal worldview.

ALFRED NORTH WHITEHEAD'S ONTOLOGICAL PRINCIPLE AND ONTIC EXPERIENCE

Alfred North Whitehead (trans. Griffin & Sherburne, 1978) explains 'concrescence' as a process within the ontological domain of experiencing. In his words, all processes being premised on continuous interrelationships between the diverse entities of various world-systems are novelties. Whitehead sees God as a necessary 'accident' in the series of novelties, all of which are considered as accidents in the sense of newness to acquire fresh learning *ad infinitum*. Thus Whitehead separates the understanding of God between the process organism concepts of some kind of metaphysics from the primordial premise of God as the ultimate 'Absolute' in other ones. The problem transmitted from either the disparateness of the primordial nature of the divine origin in these two forms, and the relations of these different premises with the world of experience, is a deeply troubling one. It reverts back Whitehead's analysis to nature by the whims of rationalism and causes all the problems for the project on unity of knowledge as were mentioned in the case of Kant, Hume, Husserl and all other Occidental philosophy of science.

The meaning of 'concrescence' as process, in view of the two disparate functions of the primordial episteme, renders the project on the possibility of acquiring knowledge solely to the ontological field. Whitehead explains the idea of concrescence that devolves ultimately to the unity of individual self (individuality) in reference to its own ego (prehension) of unity. Whitehead defines in his words (op cit. p. 211): "'Concrescence' is the name for the process

139

in which the universe of many things acquires an individual unity in a determinate relegation of each item of the 'many' to its subordination in the constitution of the novel 'one'."

The ontological domain, which in our understanding of the phenomenology of unity of knowledge, is the causation immediately following the determination of the epistemic primordial in a unique and universal precept, is changed by Whitehead into many actual entities (prehensions) that determine various concrescent entities of the experiential world-systems. This renders analogical thought into a problem of causal relations between 'one' (primordial unity of individual egos) and 'many' (the consequential experiences premised on the prehensions of individual egos).

Whitehead's (op cit. p. 19) pronouncement is, "The ontological principle can be summarized as: no actual entity, then no reason". This idea leads into the same problem of heteronomy as of Kant's. For in Kant as in Whitehead, God is relationally independent of the experiential world by his absoluteness and non-process dimensions of ultimate entity.

Contrarily, we are claiming by epistemology and formalizing by ontology and ontic projection of the entire project of phenomenology of unity of knowledge that, the relational epistemology of God (unity of divine knowledge) in world-systems is possible, and is profoundly powerful and ultimately determinable. This kind of formalism of the project of phenomenology of unity of knowledge is our road towards the universality and uniqueness of the neurocybernetic and system methodology for understanding reality and computing the quantitative nature of experience within the framework of unity of knowledge. The independence of God as the Supreme Law Giver and origin of the entire phenomenology of unity of knowledge is maintained. It is only this: The divine laws are independent of the world-systems as causation. But the world-systems do generate and reflect the necessary field for progressively fathoming the depths of the divine laws for learning world-systems according to their evolutionary novelty of precepts, formalism and evidence.

MARTIN HEIDEGGER'S PROJECT ON PHENOMENOLOGY AND THE SCOPE FOR UNITY OF KNOWLEDGE IN IT

The question of the primordial ego called as the reality of being, Dasein in Kant's works, was superbly investigated by Martin Heidegger (trans. Hofstadter, 1988). Heidegger's ideas on phenomenology of knowledge need to be understood in reference to the theory of extant order (extension). That is, bodies are claimed not to exist by themselves as they are, but only in relation to other entities. The question then to be asked is this: If such relations are fundamental to the existence of entities in extant, what then is the nature of the primal entity, upon which all entities depend in an ontological way? This is the problem we referred to in discussing the ultimate relegation of all egos to the experiential order, since the 'originary' metaphysics of God remains elusive to the senses.

The second question is this: If the theory of extant is fundamental in the idea of Dasein as the relational ego, then how is unity of knowledge impossible in all of phenomenology that has impinged upon the Occidental world-systems, such as of liberalism, utilitarianism and neoclassicism, modernism, postmodernism etc?

The definition of the primal entity, Whitehead's 'actual entity', Kant's noumena and Hume's ontological premise of knowledge, are essences of experiential things, though the authors place high stature to God in human actuality. Yet God is not functional, as it is the primordial Ego independent of all experience. Hence the knowledge of God is not knowable, and hence not transmittable to experience. The noumena and the ontological principle of the 'actual' functional entity thereby commences from the world of experience. The essence of extant in relational entities thus extends to the realm of experience without there being well-defined causality between the divine origin of reality and the experiential world-systems.

The causality here can be charted out in two ways. First, there is the causality from the source of God to the World. Then there must also be the process of knowledge formation from the World to know God through continuously evolving knowledge-flows and their evolutionary interrelationships between ontological and ontic entities and the divine laws, until such laws are fully comprehended. This final experience is realized in the Hereafter, the Supreme (Optimal) Event of Completion of all Knowledge and the ultimate realization of all world-systems.

The Occidental philosophers' view of reality does not allow for this completeness of knowledge to be known. Even if the possibility exists in the large scale universe of completion of life in its entirety in all the universes, such metaphysical meanings are found to be void of human experience. The phenomenological worldview of unity of knowledge on which we have embarked to explain reality, and then to provide meaningful and explainable relations to the embedded entities, embraces the possibility for continuity of the process relations. This expands over time and continuums of knowledge-flows from and towards the finality of divine completeness. More on this will be examined in subsequent chapters.

Thus, as we will study in a formal way in subsequent chapters, the causality generated and put into extensions of relations between entities being limited to human experience, is also limited within their scope of ontological and ontical frameworks. The epistemic origin of prehension (the Ego) premised in the divine laws, and the functional relations of this law with experience gained through co-evolutionary circular causation, do not belong to the project of Occidental philosophy of science. Hence the Dasein as the origin of reason is confined to the materially envisioned worldview becomes a constructed and anthropologically evolved reasoning process in world-systems. Entities are extant by such a vision of reality premised on a limited meaning of knowledge.

On the theme of Kantian theory of extant (extension) as being limited within the Dasein of original experience, which is the transcendental philosophy, Heidegger (op cit, p. 127) writes,

> The "I-think", says Kant, must be able to accompany all my representations, that is every cogitare and cogitate. This statement is not to be taken, however, as though the idea of the ego is present along with every comportment, with every thinking in the broadest sense. Instead, I am conscious of the linkage of all comportments with my ego, that is to say, I am conscious of them in their multiplicity as of my unity, which has its ground in my egohood (as subjective) as such. It is only on the basis of "I-think" that my manifold can be given to me. In a summary way, Kant interprets the ego as the 'original synthetic unity of apperception" What does this mean? The ego is the original ground of the unity of the manifold of its determinations in this sense, that as ego I have them all together with regard to myself.

It is true though that only the self can comprehend any reality that it is capable of. The question though is not what 'I' comprehends, but rather what makes 'I' comprehend as it does. The ultimate *a priori* seat of all knowledge emanates from the source and meaning of such comprehension. The "I-think" is a spontaneous result of such ultimate *a priori* meaning.

The quest for and answer to this fundamental problem of the relationship between the ultimate source of the meaning and comprehension of knowledge, and its projection onto the extant reality of the being of beings, that we must seek the answer of reality. We must then derive the methodology, the methods and the tools and processes of application by laws, institutions, policies and the quest of the ego for computing aspects of the constructed meaning of reality. We would then take away the centricity of self as ego, the Dasein, and instead bring it within the manifold of self and other. We would then have given meaning to the theory of extant in the perspective of the guidance of something that can be proved as universal and unique, though diverse, but not individually intentioned as ego so as to be different and multiple across over-determined domains of episteme.

The second question nonetheless is to address the meaning of individualism and individuality in the context of the theory of extant. In other words, how can we understand the following paraphrase (Heidegger op cit. p. 148): "It should be noted that finite substances, things [*Sachen*] as well as persons, are not simply extant in any arbitrary way, but exist in reciprocity, in a commercium. This reciprocal action is founded on causality, which Kant takes to be the faculty of producing effects." In reference to our foregoing argumentation the 'commercium' in the production of knowledge is designated to the ultimate ego of individuality, which when set free under its own prehension of 'actuality' determines a being in its own individuated light. The personification of all such freely determined self and ego by preferences causes the problem of over-determination, as was pointed out by Marx (Resnick & Wolff, 1987). It is also the linearly additive representation of utilitarian formalism within methodological individualism despite the presence of an ethical preceptor in utilitarian decision-making (Hammond, 1987; Quinton, 1989).

The problem is not due to a failing in the good intentions of the moral philosophers. The matter is one of methodology to discern the functional nature of the ultimate reality, which is the unique and universal order of reality, with the relations of world-systems. Consequent upon this determination rests the moral and ethical self-actualization, and the theory of extant order in relation to the functional nature of such ultimate law and guidance. Thus the answer to the question of disparate implications between normative theory and positive applications of moral philosophy is the inadequacy of Occidental moral philosophy. It is grounded on the impossibility of the Occidental Ego to rest in the divine will, and thereby to find its projection in the world-systems and their entity relations via the abiding divine laws. The theory of extant relations between entities in Occidental moral philosophy is a matter of the mutated and often conflicting nature of such relations, rather than one premised on unity of knowledge by transformation of self within the whole in the light of the universal.

Such representations between self and other in the relational commercium mentioned above, project equally to the social, economic, institutional and global orders. This subject was covered in earlier chapters. In this chapter we can advance that same coverage by the perspective on institutionalism from opposite viewpoints of the phenomenology of moral philosophy.

INSTITUTIONALISM AND PHENOMENOLOGY

Morality, ethics and values are central edicts of moral philosophy and part and parcel of the study of phenomenology. Our critical examination of modernity, post-modernity, liberalism and economic neoclassicism has led us to believe on the need for a fresh interpretation of the world in the light of unity of knowledge. This invokes also a fresh understanding of central themes that comprise humanity. They comprise issues of social justice, markets, governance and sustainability. Within this complex of central themes, the meaning of Computing Reality takes up particular shape and form. Now the IIE-process worldview that was briefly mentioned before carries along its epistemology, ontology and ontic domains in establishing the meaning and application of the codetermined world-systems between entities and their relations in them.

Sustainability itself is defined by the continuity of justice, balance, moral, ethical and material acquisitions that together configure the human social order, markets, institutions, and the global political economy. Such a complex of interrelations, which in the IIE-process methodology is formally explained by circular causation and pervasive complementarities as the sure signs of unity of knowledge, become the ingredients of human behavior, knowledge and organization of life and thought. Institutions are borne and sustained by such pervasive and continuous regeneration of dynamic preferences interconnecting self with the other across social wholes.

Computing reality in such a social and institutional framework is influenced by the methodology that emanates from the contrasting worldviews of liberalism, methodological individualism, modernism and postmodernism on the one hand. On the other hand, the IIE-process methodology of unified world-systems bestows an altogether different worldview. The IIE-process methodology establishes an interconnected normative and positive conception, bringing forth its influence in the construction of world-systems, their entities and interrelationships.

Before we bring forth the relevance of Computing Reality in all of the philosophical argumentation contra sustainability explained by the project of phenomenology of unity of knowledge, we first proceed along the following stages. Our building blocks comprise the explication of the meaning of justice, organization and economic and social constructs that lead to a theory of institutionalism. Such an explication will comprise the theory of institutionalism according to the IIE-process methodology (Choudhury, 1994). It is argued to form the unique and universal premise of sustainable human future by virtue of its universally complementary and participatory essences.

Institutionalism

In an earlier chapter we have inquired on the idea of organization and institution. To re-examine that definition let us consider the following points: Organizations are internal to institutions. Institutions are fired and driven forward by goals and ends. These derive their staying and accomplishing power from underlying episteme of life, intellection and organization of these. Feiwel (1987, p. 53) writes on Arrow's conception of organization and institution:

> An organization is a group of individuals seeking to achieve some common goals, or, in different language, to maximize an objective function. Each member has objectives of his own, in general not coincident with those of the organization. Each member also has some range of decisions to make within limits set partly by the environment external to the organization and partly by the decisions of other members. Finally some but not all observations about the workings of the organization and about the external world are communicated from one member to another.

Institutions are implementing arms of policies and programs. These are visible power structures entrusted to carry on the job of organizing and implementing policies into the socio-economic system. Yet in terms of the IIE-methodology, the intrinsic knowledge-centered IIE-process is instrumental in the first place, to empower institutions with the foundations and authority of the common will. Thus institutions are participatory in nature and are decentralized within their own specific structures. The connection between polity and institutions in respect to power and organizational behavior comprises a pure praxis of thinking and of deliberating on rules based on the praxis.

Polity, institutions and organizations of the IIE-process genre interact to simulate rather than maximize the commonly interacted and complemented dynamic goals of the common weal. This is a perspective that makes substantive difference from the nature of Arrow's organizations. Consequently, preferences formed in these two different perspectives lead to different consequences on the total meaning of human sustainability vis-à-vis institutionalism.

The resulting nature of markets, economy, society and organization is one of self-interest resting on individual welfare functions somehow aggregated across disparate goals. The maximizing goal of such aggregated social welfare function leads into an organizational structure that is internally of the competing type, and thereby consequentially of the same type for the institutions and polity that reflect on organizations. Preferences formed by the citizens of Arrow's organizations, institutions and polity are of the neoclassical orientation in self-seeking and competing behavior for the purposes of satisfying the optimal conditions of welfare maximization.

All these properties of Arrow's organizational behavior are contrary to the participatory, learning and simulating preferences of polity, institutions and organizations in IIE-process oriented framework. The limits of organizational, institutional and polity behavior are caused by a dictator in Arrow's social choice and social welfare function. Such limits are caused by the limits of the ultimate and most irreducible bound of unity of knowledge in IIE-framework. In the latter case, co-evolution of interactive and integrative processes across processes of unification (consensus) across systems, take place in perpetuity and across continuums of unifying systems. The polity, institutions and organizations are premised on evolutionary and learning preferences that continuously give shape and form to learning entities under the guidance of laws and rules that foster socio-economic, moral and ethical complementarities.

Social Justice and Economic Efficiency

Contrasting forms of preferences emerge between Arrow's social choice theory, and thereby, neo-liberal forms of polity, institutions and organizations on the one hand, and the participatory forms of the same in the case of IIE-framework. The result is specific ideas of social justice and economic efficiency concepts and their interrelationships. Two other examples need to be examined here in the liberal design of institutionalism to establish the pervasive contrariness between this and the IIE-worldview of unity of knowledge respecting the issues of institutionalism.

148

In Arrow's social choice and social welfare analysis decisions are made by individualistic preferences that are somehow aggregated together, as for instance, by the method of utilitarianism (Hammond, op cit; Harsanyi, 1955). Arrow does not want to ignore the importance of altruism in exchange mechanism (Arrow, 1974, p. 19-29). Yet endogenous learning remains absent in Arrow's social organizations and social contract due to the fact that Government is made to fill in the social gap left by the failure of the price system.

Arrow's formulation of learning curves in average cost relation is devoid of process explanation. Only after-effects of learning-by-doing are used as explanatory factors in the average cost relation (Arrow, 1962). Thereby, ethics and values remain exogenous to social choices. This is contrary to dynamic preference formation that establishes circular causality between learning entities of the economic and social systems. Consequently, preferences are linearly aggregated and the assumption of linear independence is retained in Arrow's aggregation process despite an exogenous role of social cohesion through liberalism and altruism.

Rawls' (1971) conception of social justice invokes his Difference Principle that states that social welfare is increased only by improvement of the most disadvantaged in society. Economic and social inequality is reformed by putting society behind the Veil of Ignorance. This is a Kantian representation of equality by virtue of assuming that mankind is capable of moral responsibility. Yet it remains a mystery how a non-interacting state of the Veil of Ignorance with perfectly altruistic beings in insulated compartments of thinking can be made to learn with the external world-systems and to reform them into social equality. The tenets of policy and laws and rules once again become enforced guidelines. Hence these attributes remain exogenous to the learning process found in the IIE-process oriented worldview.

Thereby, it is equally difficult to define the evolution of sequences of veils of ignorance from one level of inequality-to-equality to subsequent levels of inequality-to-equality. Questions asked are: What causes the Difference Principle to enter the mindspace of the non-interacting perfectly altruistic agents of society? Rawlsian idealism is rejected by the endogenous treatment of morality, ethics, values and preferences in the IIE-process oriented worldview and methodology. Rawlsian social and ethical aggregation reflects a utilitarian second-best position of exogenous application of laws, policies and rules being enforced in human preferences to put them in order. Rawlsian institutions thereby become exogenously defined, as in Arrow's case of social choice, to reaffirm the continuing saga of neo-liberalism. This is to erect a social order of linearly aggregative preferences, while missing out the otherwise complex nature of social aggregation in reality.

On the last point, we note the incisive description by Rousseau (trans. Cranston, 1968) in how linear aggregation of independently constituted individualistic preferences of ideal types is tantamount to utterly meaningless social wholes. There is likewise the utterly meaningless idea of social deconstruction. On both the critique of linear aggregation and linear deconstruction Rousseau writes (p. 71, edited): "It is said that Japanese mountebanks can cut up a child under the eyes of spectators, throw the different pieces into the air, and then make the child come down, alive and all of a piece. This is more or less the trick that our political theorists (*economic theorists*) perform – after dismembering the social body with a sleight of hand worthy of the fairground they put the pieces together again anyhow."

Like Rousseau, Gunnar Myrdal comes nearer to discard the segmented view of social and economic systems caused by linearity and individuation. Myrdal's conception of social problem is contrary to Arrow's and the neoclassicists in that it makes no such differentiation in the understanding of holistic social problems. He takes them all as an ensemble of problems interacted with each other. Myrdal (1979) writes: "... in reality there are no economic, sociological, psychological problems, but just problems and they are all mixed and composite." Such an integrative view of problem solving led Myrdal to criticize the neoclassical economic reasoning of rationality, methodological individualism and independence. On this issue Myrdal (1987, p. 274) wrote: "Hundreds of books and articles are produced every year on "welfare economics", reasoning in terms of individual or social 'utility' or some substitute of that term. But if the approach is not entirely meaningless, it has a meaning only in terms of a forlorn hedonistic psychology, and a utilitarian moral philosophy built upon that psychology. I have always wondered why the psychologists and philosophers have left the economists alone and undisturbed in their futile exercise."

In the end, we note that precepts of social justice, fairness, economic efficiency and material progress, morality, ethics and values, must all rest on a systemic view of social interaction. Earlier we explained this in terms of embedded systems as opposed to differentiated systems. The goal of sustainability can be configured, pursued and attained as simulated learning from lesser to higher degrees of knowing in the embedded relations of unified systems. Contrarily, the same precepts become human perceptions premised on the original point of the rationalistic Dasein, and what this entirely means in the moral philosophy of the Occidental order.

Upon such precepts rest equally the institutionalism of Occidental genre in contradistinction to the learning, participatory and pervasively complementary world-systems of the IIE-framework. In subsequent chapters we will identify the various instruments and rules that make the IIE-perspective of social justice, economic progress and sustainability functionally attainable along the simulated path of pervasive learning.

In the context of the phenomenology of unity of knowledge, the neurocybernetic and systemic understanding of institutionalism within the inquiry of moral philosophy, and spanning the socio-scientific order, now brings us to the role of all of these in the project of Computing Reality. It is to be noted here that the human order, interspersed as it is by diverse and multidimensional aspects of values and material forces, institutionalism and its driving social preferences, laws and rules, presents a complex system. In such a system only, order out of complexity prevails. The neurocybernetic perspective is to study the circular causality of evolutionary learning processes between the embedded systems and their entities. All the formal rules and applications emerge thereby for designing and sustaining a fresh promise to human future.

The assumption is thus made that there is always hope for man. This optimism is quite contrary to what we note in many mainstream economic and social expressions. For instance Heilbroner (1991, p. 20) sounds the parting knell of human future: "Thus, to anticipate the conclusions of our inquiry, the answer to whether we can conceive of the future other than as a continuation of the darkness, cruelty, and disorder of the past seems to me to be no; and to the question of whether worse impends, yes."

On the sustainability question too, system failure caused by irreversible entropy, such as the one exemplified by the arrow of time and the build-up of entropy conditions, is seen not to be a permanent condition in the IIE-lens. We now turn to these issues in the light of neurocybernetic and system theory as seen from the side of computing reality premised on the phenomenology of unity of knowledge.

COMPUTING REALITY IN MORAL PHILOSOPHY MODEL

Consider the following structure of deconstructions that are interdependent between them by virtue of the relational epistemology that defines unity of knowledge. Such characteristics as mentioned below can and are inter-related within their own groups. Consequently, their continuity is maintained across continuums of space and time. Such is also the implication of the theory of extension underlying each of these groups differently and oppositely, as shown below:

Moral Philosophy

Reductionism to Rationalism	Reductionism to Unity of Knowledge
1. Individuality and individualism = Dasein	1. Individuality within the collective = Learning on unity of knowledge
2. Mutation of groups	2. Pervasive participation and complementarities across time and continuums
3. Civilization Consequences: Liberalism, Modernity, Post-Modernity	3. Post-modernist reconstruction
4. Nature of individualistic social morality, ethics and preferences	4. Endogenous learning preferences according to the IIE-process methodology: unity of knowledge
5. Institutionalism premised on political economy of pricing, markets and competition	5. Institutionalism on globally participatory social contracts

Now note that the presence of individualism as the groundwork of socio-scientific rationality in the rationalistic domain results in the impossibility of interaction between preferences and agents. Such is the characteristic of all linear systems in terms of absence of interaction, which are otherwise caused endogenously by learning between entities, as shown. The absence of learning relations caused by interaction, integration and dynamic evolution in the Rationalist Order results in the impossibility of a neurocybernetic and system theory of unity of knowledge. Just the opposite is true of the phenomenology of unity of knowledge. Here linear non-learning behavior is a degenerate case of IIE-process methodology. The prevalent nature though is of non-linear dynamics of learning in continuity and across continuums. Thus all entities in such an order must necessarily be influenced by perpetuation of the same kind of behavior. Now neurocybernetic and system theory becomes a natural result of phenomenology of unity of knowledge.

Within the above form of embedded ensemble of organisms against rationalist deconstructions in linear non-interacting relations, computing reality becomes a study on reflective empiricism. That is, the combination of epistemology, ontology and ontic consequences results in the form of quantitative and empirical methods that, fall back upon the IIE-process methodology. Thereby, the methods that emanate from this conscious background of the study of phenomenology are those that comply with and in turn reinforce the methodology of the corresponding phenomenological worldview.

An example here is that simulation as a method replaces the method of optimization in the case of the linear non-learning spaces. The method of optimization now remains as a background for criticism by that of dialectics in establishing the meaningfulness of the simulation method only in learning systems. Another example is the primacy of knowledge over time as the foundational groundwork of change. Time acts as a recorder of such change but not the cause of change. The neoclassical economic ideas of marginal rate of substitutions, competition and full information do not exist, except as ideas for criticism in the reflective empiricism of computing reality of unity of knowledge.

Finally, we note that between the two cases – Rationalism and Unity of Knowledge, there are no interaction except that Rationalism and its emanating methods become objects for rejection by the dialectics of phenomenology of unity of knowledge. Thereby, the building blocks of a learning neurocybernetic and system theory of unity of knowledge remain unique and universally distinct in their own specialized nature. We also note that learning neurocybernetic and system theory carries the methodology of unity of knowledge not simply across the human order but also in the domain of machines (computation). This is so, because the carrier here is consciousness of the background methodology that establishes the learning mindspace. Such a mindspace then derives methods on the basis of its background methodology. Application follows on the basis of the ontic consequences.

Machines like humans are prototypes of learning. Hence computing is a way of enabling as well as preserving the organizational, measurement and applied perspectives of the methodology of learning in unity of knowledge. Machines though do not reproduce themselves as an organism. This is not the meaning of learning pertaining to machines. Rather, in the neurocybernetic and systems theoretic sense such learning becomes the circular cause and effect of a perpetuating unique and universal praxis that underlies all socio-scientific worlds.

Thus, machines are regenerated and continued across evolutionary paths by their improvements and enhancements with greater depths of understanding along the evolutionary nature of the learning methodology of unity of knowledge. In this way, machines are like conscious phenomenological artifacts in the pursuit and application of an ever-broadening and deepening understanding of the learning methodology as pronounced by the IIE-process methodology. This involves spontaneity between epistemological, ontological and ontic phases of knowing.

Wiener (1966, p. 30) writes in the above context: "However, different as the mechanical and the biological reproduction may be, they are parallel processes, achieving similar results; and an account of the one may well produce relevant suggestions in the study of the other." Subsequent chapters will dwell more formally in establishing such a unique and universal methodology of the formal logic of unity of knowledge.

CONCLUSION

We have discussed the theory of phenomenology as the study of consciousness by invoking the epistemology and ontology of unity of knowledge in cybernetic and system theory, human systems and machine theory. Thus the theory of unity of knowledge embedded in neurocybernetic and system theory has been established as a unique and universal praxis of thought. The theory of institutionalism in respect of unity of knowledge as a relational world-system is contrasted from linear ways of learning. Complexity and non-linearity are shown to be extant in the theory of unity of knowledge as it pervades the phenomenological worldview of mind, matter and machines. Linearity is not capable of explaining such complexity that arises by richness of learning relations.

The project of phenomenology within the epistemological, ontological and ontic concepts is given a new and original understanding that remains different from metaphysical unity and brings the divine order nearer to human experience. The engineering meaning of ontology is thus used and is established in relation to the primordial role of epistemology and its ending point of ontic evidences in the interactive, integrative and evolutionary model of unity of knowledge in relational processes.

REFERENCES

Arrow, K.J. 1951. *Social Choice and Individual Values*, New York, NY: John Wiley & Sons.

Arrow, K.J. 1962. "The economic implications of learning by doing", *Review of Economic Studies*, Vol. 29, pp. 155-173.

Arrow, K.J. 1974. *The Limits of Organization*, New York, NY: W.W. Norton.

Carnap, R. 1966. "Kant's synthetic *a priori*", in his *Philosophical Foundations of Physics*, Ed. M. Gardner, New York: Basic Books, Inc.

Choudhury, M.A. 1994. *Economic Theory and Social Institutions*, Lanham, Maryland: University Press of America.

Choudhury, M.A. 1997. "The epistemologies of Ghazali, Kant and the alternative: formalism in unification of knowledge applied to the concepts of markets and sustainability", in J.C. O'Brien Ed. Special Issue of *International Journal of Social Economics*, Vol. 24, Nos 7/8/9: *Essays in Honour of Clement Allan Tisdell Part III*, 1997.

Choudhury, M.A. 1998. "A Socio-Scientific Theory of Continuous Machines", *Cybernetica* Vol. XL!, Nos.2/3/4.

Dawkins, R. 1976. *The Selfish Gene*, New York: Oxford University Press.

The New Encyclopedia Britannica, Macropedia Vol. 14, 1981. "Phenomenology".

Feiwel, G.R. 1987. "The many dimensions of Kenneth J. Arrow", in Feiwel, G.R. ed. *Arrow and the Foundations of the Theory of Economic Policy*, London, Eng: Macmillan.

Hammond, M. Howarth, J. & Keat, R. 1991. *Understanding Phenomenology*, Oxford, UK: Basil Blackwell.

Hammond, P.J. 1989. "On reconciling Arrow's theory of social choice with Harsanyi's Fundamental Utilitarianism", in G.R. Feiwel Ed. *Arrow and the Foundation of the Theory of Economic Policy*, pp. 179-221, London, Eng: Macmillan.

Harsanyi, J.C. 1955. "Cardinal welfare, individualistic ethics, and interpersonal comparison of utility", *Journal of Political Economy*, 63: 309-321.

Heidegger, M. trans. Hofstadter, A. 1988. *The Basic Problems of Phenomenology*, Bloomington, IN: Indiana University Press.

Heilbroner, R. 1991. *An Inquiry into the Human Prospect*, New York, NY: W.W. Norton.

Hume, D. reprinted 1992. "Of the understanding", in his *Treatise of Human Nature*, Buffalo, NY: Prometheus Books.

Husserl, E. trans. Alston, W.P & Nakhnikian, G. 1964. *The Idea of Phenomenology*, The Hague: Martinus Nijhoff.

Kant, I. trans. Paton, H.J. 1948. *The Groundwork of the Metaphysic of Morals*, London, Eng: Hutchinson.

Kant, I. 1949. "Critique of pure reason", in Friedrich, C.J. ed. *The Philosophy of Kant*, New York, NY: Modern Library.

Kant, I. trans. L. Infeld. 1963. "The general principle of morality"; also see "Natural religion", "Prayer", in *Kant's Lectures on Ethics*, Indianapolis, IN: Hackett Publishing Co. 1963.

Myrdal. G. 1979. "Institutional economics", in his *Essays and Lectures After 1975*, Kyoto, Japan: Keibunsha.

Myrdal, G. 1987. "Utilitarianism and modern economics", in Feiwel, G.R. *Arrow and the Foundations of the Theory of Economic Policy*, London, Eng: Macmillan, pp. 273-278.

Quinton, A. *Utilitarian Ethics*, La Salle, IL: Open Court, 1989.

Rawls. J. 1971. *A Theory of Justice*, Cambridge, Massachusetts: Harvard University Press.

Resnick, S.A. & Wolff, R.D. 1987. *Knowledge and Class, A Marxian Critique of Political Economy*, Chicago, ILL: The University of Chicago Press.

Rousseau, J-J trans. Cranston, M. 1968. *The Social Contract*, London, Eng: Penguin Books.

Russell, B. 1990. "Bergson", in his *A History of Western Philosophy*, London, Eng: Unwin Paperbacks.

Sztompka, P. 1991. *Society in Action, the Theory of Social Becoming*, Chicago, IL: The University of Chicago Press.

Whitehead, A.N. trans. Griffin, D.R. & Sherburne, D.W. 1978. *Process and Reality*, New York, NY: The Free Press.

Wiener, N. 1961. *Cybernetics*, Cambridge, Massachusetts: The MIT Press.

Wiener, N. 1966. *God and Golem, Inc*, Cambridge, Massachusetts: The MIT Press.

Wolff, R.P. 1977. *A Reconstruction of a Critique of a Theory of Justice: Understanding Rawls*, Princeton, NJ: Princeton University Press.

CHAPTER 6: NEUROCYBERNETIC AND UNITY OF KNOWLEDGE ACCORDING TO THE SCHOLASTIC THINKER ABU HAMID Al- GHAZALI (1058-1111)

In our earlier chapters we have construed the inner philosophical and applicative meaning of knowledge and how it constructs unified world-systems according to the episteme of unity of knowledge. Contrary to this kind of a worldview was shown the worldview of individuation, marginalism conflict and segmentation. This was true of mainstream economic reasoning as organized under neo-liberalism and the entire gamut of the world-system premised on competition, scarcity of resources and knowledge-benign decision-making and behavioral preferences. Yet the structure of scientific arguments in the framework of neurocybernetic learning systems premised on the epistemology of unity of knowledge presents a world-system quite contrary to this. Thereby, deep philosophical, scientific and ecological questions enter the framework of neurocybernetic reasoning.

The great scholastic thinkers thought of knowledge and reasoned its role in understanding physical, social and intrinsic reality within the framework of an interactive and integrated design of the universe. Such an embedded way of understanding reality is similar to the present days approach in the study of systems and cybernetics using learning formalism. But the scholastic masters thought much more on a cosmological design of reality than on formalism underlying such reality. Thus their approach to the systems view of the universe and its learning entities was much more in terms of moral and ethical values that remain intrinsic to existence.

THE PHENOMENOLOGY OF ABU HAMID AL-GHAZALI

Abu Hamid [Imam] Al-Ghazali (1058-1111) was one such scholastic thinker in the Islamic traditions who lived before the Eighteenth Century European Enlightenment. His scholarly works were transmitted to Europe and formed a basis of the scholasticism of Thomas Aquinas, Albert the Great, Roger Bacon and others of that age. In this regard Buchman (1998, p. xxii) writes [edited]: "The work was so clear and presented such a knowledgeable summary of Peripatetic philosophy that, when it was made available in Latin translation to the Christian scholastics, Albert the Great, Thomas Aquinas and Roger Bacon all repeatedly mentioned the name of the author of the 'Intentions of the Philosophers' along with Ibn Sina and Ibn Rushd as the true representations of Arab Aristotelianism."

The philosophical and psychological groundwork of a neurocybernetic project in unity of knowledge in which Imam Ghazali excelled was premised on the precept of Oneness of God (Tawhid in the Qur'an). Through the Tawhid-centered worldview Ghazali explained the union of self and the artifacts of life in a manner that provided universality and character that was equally acceptable to Europe and to the Muslim World at that time. But speaking of today's climate of understanding in the trans-empirical nature of holistic system theory and neurocybernetic, the message that Ghazali left behind was of a singular nature.

Modern system and cybernetic theory does not premise its foundations on the holistic human nature and its application in comprehending reality. In this regard Buchman (p. xviii) writes "People today – Muslims and non-Muslims alike – usually hold drastically different assumptions on the nature of existence than those held by al-Ghazali and most of his contemporaries. Knowing the tacit but prevalent presuppositions of the medieval Muslims *Weltanschauung* enables the contemporary reader to understand the depth and beauty of al-Ghazali's interpretation of divine unity."

Yet Ghazali did not practice sheer Gnosticism. He combined his mystical insight in the craving and ascent towards God and the knowledge of divine unity bestowed on his thoughtful and deepening mind with application of such purity to practical affairs of life. In the former case he was a mystical thinker. Yet this was not tantamount to scholastic rationalism as was the case with the contemporaries of his time, such as Ibn Sina, Ibn Rushd, Al-Kindi and Al-Farabi (Qadri, 1988). The rationalists thought that the understanding of reality in its complex form of interrelationships was affordable through reason. Any permanent reference to the divine texts and thus to the epistemology of oneness of knowledge premised on the fundamental Islamic sources, the Qur'an and the Sunnah, was not necessary. To Imam Ghazali reason was not enough to find the inner depth of the divine meaning embalmed in reality. The quest and knowledge of such trans-empirical experiences combining self with the world through God depended principally on the purity of self and the earnest and deepening quest for understanding the unity of the divine laws.

The second and coterminous nature of Ghazali's thought rested on the application of divine purity of self and mind to the practical artifacts of life and society. Ghazali referred this part of total knowledge to the Islamic Law (the Shari'ah). He claimed that the core of the Shari'ah depended upon the purity of mystical belief and true quest for the oneness of God. It can then be implied that the unification between self-purification and the instrumental laws of the Shari'ah enabled the foundation of a good society that was God-centered, and thereby holistic, morally embedded, rich by complexity and overarching between all disciplines of life. In regard to such integration between deeper truth and the development of the Shari'ah by Ghazali, Buchman (p. xxvi) points out, "He saw that to revive Islam among people the shari'a must be infused with the practice of its inner dimension, Sufism."

Ghazali's two integrated perspectives of human and social existence were like Adam Smith's view of human nature and the social and economic world-systems six centuries later. In his Theory of Moral Sentiments, Smith described the ideal man of empathy. On the basis of this, human behavior and society were delineated. In the Wealth of Nations, Smith delineated markets, economic behavior and institutional development. The normative law of natural liberty was applied to crown the freedom of man in society. The result was unbridled individualism of self faced with competition, scarcity and self-interest behavior in economic and social dealings. Smith's high social principles in The Theory of Moral Sentiments thus crumbled into a treatise of human competition and self-interest in the Wealth of Nations.

Such an intellectual incident was though contrary in the case of Imam Ghazali. Ghazali's model of man was permanently capable of enabling him to reach God through purity of self in the quest for unity of being. Such a belief pronounced itself in all his subsequent works combining self with the world through God. It was heightened in Ghazali's Ihya Ulum Id-Deen (Revival of Religious Thought) (Karim, undated). Ihya was the masterful text explaining how philosophy, morality and law combine with human psyche to give true meaning to worldly matters and artifacts.

Karim (p. 53) states how Ghazali understands the relationship between God-consciousness and worldly activity exemplified in a special case of economic earnings. Ghazali delineated the reactions of three kinds of men to the activities of the world (earnings). Firstly, there is the person who forgets the Hereafter, when and where mankind will be judged by God for the goodness and evil he has reaped in worldly life. Instead, the person submerges into worldly pursuits alone. Such a person will be destroyed both in life and in the Hereafter before God's justice. The second kind of person is one who treats the worldly earnings for the sole purpose of attaining the bliss of the Hereafter. Such persons pursue the true and pure path of worldly acquisitions, and thereby bring about wellbeing for all that their lives and actions touch. Such a person is the blissful winner in this world and of the next according to divine justice. The third category of persons is midway between the first two. They will receive the divine blessings as long as they remain along the middle path of divine bliss.

In such ways, Ghazali uses the medium of the Qur'an, the guidance of the Prophet Muhammad (Sunnah) and his own deep quest for truth to explain the systemic link between worldly matters and the Hereafter, which is the event of the penultimate divine judgment for eternal reward or punishment. In his Ihya, Ghazali has spanned a vast area of such worldly activities on establishing the good society in the light of the divinely guided self and other. Ghazali wrote on a long list of worldly pursuits governed by human self in abidance with the divine laws. This list included earnings, trade and commerce, rules of eating and drinking, contract of marriage, earnings, trade and commerce, permissible and forbidden acquisition, social harmony, kinship duties as according to the Qur'an and the Sunnah, acquiring of knowledge as a supreme value of the believer, and many other ones. The embedding of all such worldly actions in response to and by establishing circular causation between moral values and such worldly artifacts defines the spiritual definition of goods, services and all forms of resources. From such morally transformed artifacts and desires arise the ethical nature of market exchange, capital formation and distribution of resources, wealth and opportunities to gain social wellbeing. Nothing that is worthy of the true desire of worldly things embedded in the divine laws and guidance is left out from blissful transformation.

Ghazali notes the ascent of humankind through stages of devotion to and quest for God. Along with these ascending heights of divine accomplishment are realized the increasingly better organization of life and thought. Ghazali writes on such coterminous advance of the human moral worth and the divine blessings in various ways. In his Ihya, Ghazali writes (Karim, p. 245, Vol. 4, edited): "When the pursuer (the empirical and trans-empirical observer) in the path of religion heard about this knowledge he knew of his defects and was enraged at his passions and his mental fire broke out. Before this the light in his heart was dimly burning, even though it did not touch the fire. When knowledge was puffed up in his heart, his oil was enkindled. Then light upon light came to him. Then knowledge said to him: Value this moment greatly. Open your eyes, so that you may find the path. When he opened his eyes, he found the pen of God as described. It was not made of reed, it has no head. It is incessantly writing in the mind or soul of men. He said being surprised at it: What a good thing is knowledge. I don't consider this pen as that of the material world."

Ghazali's System and Cybernetic Thought in his *Niche of Lights*

Furthermore, Ghazali thought profoundly on the nature of knowledge and its integrative role in moral self-actualization. In such epistemological and ontological thinking Ghazali was a distinctive forerunner of Immanuel Kant and his contemporaries of the European Enlightenment. Ghazali's Niche of Lights (Mishkat al-Anwar) is a sublime exegesis of the Qur'anic verses comprising Chapter 24, verses 35-40 (trans. A. Yusuf Ali, 1946). In this, Ghazali presents a delineation of the divine origin and guidance of knowledge, whereby the limitations posed by rationalism in Kant and his contemporaries of the Enlightenment, disappear. A model of harmonious and continuous interrelationship between all knowledge-induced actions and responses is established. Thus a spiritually embedded world-system is premised on the oneness of God (Tawhid).

The Qur'anic exegesis of "God is the Light of the Heavens and the Earth" (Qur'an 24:35) is that of the divine laws, which is pervasive across continuity of time and continuums of domains creating an incessant inducement of all empirical

166

and trans-empirical entities and relations by Tawhid. Nothing escapes the divine laws and will. Consequently, nothing in the universe can escape the functional reality of the divine laws. Thereby, when the divine laws touch all entities across knowledge, time and space, it is impossible for the divine laws (God) to remain away from any activity. The problem of heteronomy that divided Kant's Pure Reason from Practical Reason does not exist in the moral law of the Qur'an, as explained by Ghazali in his Niche of Lights. The noumena and phenomena, the a priori and a posteriori, revealed knowledge and reason, all become relationally connected extensions according to Ghazali's exegesis of the above-mentioned Qur'anic verse. This belief was contrary to both the Muslim rationalists of Islamic scholasticism and those of the European Enlightenment and thereafter.

In using the concept of continuum and differentiating it with that of continuity we mean here the pervasiveness of knowledge-induced forms in the space of entities overarching continuums. An example given by Bertrand Russell (undated) is of the mathematical system as a continuum of groups of infinities pertaining to the cardinalities of the number system (Cantor, 1955). By the concept of continuity on the other hand we mean time-continuity. Since time is affected primarily by knowledge induction, and thereby records the relationship of entities to knowledge over continuums, therefore time has a borrowed character of continuity that is primarily established by the primal existence of knowledge-flows. Yet in a subtle way, which we will explain below, knowledge, time and entities are interactively integrated and evolved by circular causation.

The consequences on moral and socio-scientific reasoning are now profoundly different between the worldview of unity of divine knowledge (Tawhid) and the rationalist worldview of matter and mind dualism with its categorical multiplicity and differentiations. In the knowledge-induced continuums the unified and learning relationships establish the process of interconnectedness between the good things of life. As for the false artifacts that depart from the primal episteme of unity of knowledge (Tawhid), these too develop a dynamics of decadence and flawed reasoning. Ghazali continues on to explain this negative reality that does not link up with goodness, giving the two categories an ultimately separated meaning and conferring separated and opposite worldviews to them.

Buchman writes (p. xxxiv) on Ghazali's exegesis of the Qur'anic verses: "People are veiled by sheer darkness, darkness mixed with light, and light alone … Hence people end up worshipping the gods of their deceived perceptions". Buchman (p. 42) continues on to explain Ghazali's belief: "The rational faculties of the unbelievers are inverted, and so are the rest of their faculties of perception, and these faculties help one another in leading them astray. Hence, a similitude of them is like a man "in a fathomless ocean covered by a wave above which is a wave above which are clouds, darkness piled one upon the other" [Qur'an 24:40]. In this way, the light, which spans all world-systems by knowledge, is kept aloof from the world-systems in which continuity of knowledge does not enter. Consequently, such opposite world-systems, being devoid of the divine laws and guidance are destined to destruction. Only the face of God, which allegorizes the divine laws, survives.

The opposite constructs of world-systems are thus laid out in the Qur'anic exegesis by Ghazali. True systems and cybernetic relations are continuous in form over knowledge, time and space; knowledge being generated out of unity of the divine laws. But in system methodology every point of an event along the path of learning in unity of knowledge, two coterminous and complementary entities exist without fail. These are namely, the state of the system that continuously interacts, integrates and evolves by its own intrinsic knowledge of systemic participation and linkages. The Qur'an refers to this phenomenon of continuous and pervasive complementarities as the universe designed in pairs that reflect unity of the divine laws impacting upon every paired entity of world-systems. The thoughtful take

168

contemplative wisdom (Fikr) from such natural manifestations both in the realm of the empirical (observable = Shahadah) entities and that of the trans-empirical (Ghayb) domain where only mathematical relations exist, combining analytical philosophy explaining unity of knowledge.

Ghazali thought of such reflective events in the learning universe under the attribute of his conscious devotion to God as the moment of reading both of the above-mentioned specifications of any phenomenon conjointly. That is, the logical formalism of unity of the divine laws prevalent as intrinsic property of entities is combined with the institutional need for discourse, analysis by observation and consultation, giving guidance to those who seek guidance towards organizing the issues under study into a domain of consciousness and experience. In such a grand pervasively complementary vision of reality Ghazali saw the merging of personal psychology with the tenets of the Islamic Law, the Shari'ah. He found in the core of the Shari'ah the primal and immutable seat of the divine laws. Such a core remained immutable over space, time and knowledge-induced continuums, because the truth of unity of the divine laws was indelible. The periphery of the Shari'ah is discoursed by means of human consultation and given temporal interpretation. While the core of the Shari'ah remains immutable, the periphery is always in flux. In the latter domain can be effected the changes and revisions to the interpretation of the Qur'anic verses and the Sunnah on specific issues and problems under discourse.

In the context of neurocybernetic project of consciousness, the co-evolution and intermingling of the core with the periphery of the Shari'ah in respect to any issue and problem under socio-scientific examination is premised on the interactive, integrative and evolutionary methodology (IIE, mentioned in earlier chapters). The emergent learning by circular causation between such multi-systemic entities establishes the historical path of knowledge-induced trajectory of change, confirmation and movement.

Contrary to the world-systems in all its details, explained by the episteme of unity of knowledge, Ghazali explained the contrary picture of de-learning, wherein the divine light does not enter. Truth as light and falsehood as darkness thus permanently define existence and consciousness in the light of the divine will. In the appendix to this chapter a mathematical formalism is presented with respect to the polar views of reality in respect of truth and falsehood.

At this point, we note simply how Ghazali interpreted the Qur'anic verse in regard to the uncompromising nature of truth and falsehood in reference to the divine will. The Qur'anic verse (24:40) on which Ghazali wrote his exegesis is, "To whomsoever God assigns no light, no light has he". Ghazali mentions in his Niche of Lights the three categories of darkness that dim human consciousness and the capability to be objective and true, even though such persons may have some stint of truth in them by virtue of their pursuit of acquired knowledge. The case here is this. Primal knowledge of unity of the divine laws carries with it the earnest quest for understanding unity of worldly relations between entities. Likewise, there are those who inquire into such a project not from the side of God as the primal truth. Both groups seek to reach the world of unity of knowledge but from contrary episteme. The commonality of purpose bestows upon these groups of differentiated quanta of light from the divine source. The true pursuer of God's unity finds it as full light. The denier of God remains in the darkness of ignorance. The third category of pursuers of truth is mixed up between light and darkness.

In the last category are people, whose darkness grows out of their sensate way of viewing, contemplating and understanding the world and its inner relations. The light of objective knowledge in such pursuers is dimmed by rationalistic imagination and corrupt rationalistic speculation that bear only esotericism, not objectivity. Thus while the source of all life and existence is God as the One, the ontology that is reflected in the organization of reality by unity of systemic knowledge, the formalism of the inverted pursuer is deepened into de-learning universes. Such universes descend into linear forms representing methodological independence, individuality and mutations causing paucity of *inter*relations. In the neo-liberal context we saw such consequences in methodological individualism. We found the incapability of mainstream economics to answer the endogenous question of ethics, values and morality within the embedded social order. Consequently, the universally spanning system and cybernetic model of consciousness and reality is broken by the disappearance of circular causal relations between all forms of entities. Where such causal relations fail to exist there the continuity of unification of the entities by the project of unity of knowledge ceases to exist. The domain of explanation and universal extant of knowledge is thus constricted into a narrow field.

The result is that although both truth and falsehood are ordained by the divine laws as clear and distinct realities, the constricted nature of falsehood is proven by the supremacy of the truth universe. Consequently, the power of creativity and the extant of causal interrelations between entities of all kinds remain abundantly more than the relationally paired nature of the universe of unity of divine knowledge. The dimensionality of the falsehood domain descends into narrower systems of explanation as rationalism deepens into darkness. Dimensionality is represented by the number of explained events caused by unifying interrelations.

With such dual and polar existences of the two realities, light and darkness, one must reflect upon the nature of the other as opposites. We have noted earlier in this book that the project of rationalism and ego of self, as found in the intellectual tradition of the occidental world, marks that growing form of individualism and individuation of self and systems. This is true both in the agent-specific case and Darwinian mutation or any such development of an uncharted nature of evolution of the mind without an epistemological foundation. Contrarily, the episteme of conflict and competition and differentiation of systems and ideas drive the occidental intellectual order into dialectical opposites. No synthesis is possible in such a state of differentiation. Now symbiosis across systems means lateral aggregation of the relations that are individuated within differentiated systems. Finally, all these developments degenerate into independently linear forms based on the fact that there is no foundational epistemology of unity of knowledge that pervades all systems and diverse entities in such cases. The over-determination problem delineates the ultimate nature of all projects in rationalism.

A system and cybernetic model of relational world-system issues relies upon the dimensionality of the two domains. Firstly, there is the domain of unity of knowledge along with its dimensionality of the domain of knowledge-induced entities. Secondly, there are the systems that degenerate into mutations after a period of temporary learning, or simply into methodological individualism between the entities and their relations. When such ideas are imputed into Ghazali's meanings of Light and Darkness in his exegesis of the Qur'anic verses on Light and Darkness, we note that Ghazali's consciousness was contrary to the occidental philosophers of science. The most important in such a comparison was Immanuel Kant, who emulated Ghazali in his ideas of Reason and the moral imperative, and in the impossibility of acquiring objectivity through the passage of the a posteriori (sensate) world onto the a priori (intelligent) world. While Kant's problem of heteronomy made it impossible for the divine laws to be mapped onto the world-system, to Ghazali this kind of topological relation and function is essential. Yet in Ghazali's epistemology such cases are all in one. They are possible, are true but contrary to truth. Thus by conflict and indeterminism such cases represent the weakness in objectivity prevailing in the dark domains of reality.

Ghazali's Philosophical Refutation and Neurocybernetic Consciousness

One of the important features of system and cybernetic theory, especially for the case of Turing (1936) machines and artificial intelligence problems is problem-solving in continuum. In this case, the relational phenomenon can be studied in continuums across knowledge, time and the space of knowledge-induced forms. Ghazali's philosophical insight studied this kind of pervasiveness of relations caused by unity of knowledge across all spans of learning and observation. In such a case of continuity of relations caused by circular causality between reflexive forms, it is important for system and cybernetic theory to be able to discern the flow in continuity, as along a flow chart. The end is to identify the cause and effect and their circularity in problem-solving.

According to the methodology of unification and continuity the deductive and inductive logic, the a priori and a posteriori reasoning, the primacy of the divine laws and the pursuit of unity of knowledge by means of the divine guidance, bring about reflexive relations between the forward and backward linkages. This is also the methodology of interaction, integration and evolution of learning systems in unity of knowledge. Ghazali pondered over this important analytical issue of circular causality. He writes in his Incoherence of the Philosophers (Tahafat al-Falsafah) (Marmura, 1997, p. 158): "In brief, every event has a temporal cause, until the chain of causes terminates with the eternal celestial motion, where each part is a cause of another. Hence, the causes and effects in their chain terminate with the particular celestial motions. Thus, that which has a representation of the movements has a representation of their consequences and the consequences of their consequences to the end of the chain. In this way, what will happen is known. For [in the case of] everything that will happen, its occurrence is a necessary consequence of its cause, once the cause is realized. We do not know what will happen in the future only because we do not know all the causes [of the future effects]."

The possibility of complete determination of all causes and effects is only in the episteme, not in the cognitive and material being of the world. Consequently, the consciousness of Ghazali's neurocybernetic thought is to start and continue from the episteme of divine unity and form knowledge and knowledge-induced entities on the basis of the divine will. As knowledge increases in terms of the perfectly divine premise, so also do the insight into the knowledge of truth and the determination of cause and effect, and thereby, of the nature of things so induced. All these capabilities increase. The true knower is the one who is the most conscious of the divine laws.

Just as Ghazali characterized three kinds of knower of the divine bliss in decreasing paths – the true pursuer, the hardened sinner, and the blemished self, so also he placed the possibility of knowledge at three phases. In his Incoherence of the Philosophers (Marmura, p. 217) Ghazali writes (edited): "Whoever is deprived of the virtue in both moral disposition and knowledge is the one who perishes. For this reason God, exalted be He, said: "Whoever purifies it has achieved success and whoever corrupts it fails." (Qur'an 91:9-10). Whoever combines both virtues, the epistemological and the practical, is the worshipping "knower", the absolutely blissful one. Whoever has the epistemological virtue but not the practical is the knowledgeable [believing] sinner who will be tormented for a period, which [torment] will not last because his soul had been perfected through knowledge but bodily occurrences had tarnished [it] in an accidental manner opposed to the substance of the soul."

By premising consciousness and the sublime success on knowledge and epistemology, Ghazali enunciated his praxis like Einstein. Einstein too claimed that there can be no science without epistemology (Bohr, 1985). But since the epistemological-practical believer is the truly successful according to Ghazali, it implies that the relationship between the divine laws and the world is a continuous one in time over continuums of knowledge-induced universes.

Cognitions and forms that appear and disappear in the context of continuums caused by unity of divine knowledge are in turn induced by heightened levels of bliss, unraveling of vision and understanding. In every case, we find that Ghazali's consciousness was configuring the conscious artifact of the universe that continuously and consciously seeks knowledge of divine unity in the scheme of things.

The neurocybernetic analytical structure constructed on the basis of this kind of consciousness and epistemology is tantamount to the mind of the supercomputer that must think across continuums of relational entities and fields, all at once. Such an approach to complex problem-solving implies simulation of information to the extent of solving the largest domain of problems and expanding this to even higher capacity of problem-solving (Kuskov, 1999). This can be possible if and only if the law of unity of knowledge remains fixed at the primal origin and guides the entire system of causality onwards and by reflexivity into higher levels of comprehension of the episteme of unity of knowledge in the order of existence.

Knowledge, Consciousness and Event

Ghazali's epistemology of human consciousness in the light of the divine laws and its neurocybernetic implications can now be formalized into a 'BIT' that enters the systemic chain of knowledge-induced relations. The point to note at the outset is that primal knowledge and belief are coterminous premises of reality according to Ghazali's epistemology. The relationship between belief and knowledge is this. While primal knowledge is absolute, perfect and complete in the divine laws, as a perceiver the pursuer's knowledge remains permanently imperfect. The pursuer's knowledge is ever changing, either in the direction of the knowledge of divine unity or falling into decadence as knowledge tarnishes and moves away from the divine core. This latter form of perception as of the rationalist type, despite being formal and analytical in itself within its domain, is not the carrier of unity of knowledge. We will refer to it as de-knowledge in the sense of its inability to realize unity of divine knowledge. The latter alone is the ultimate premise and fact of truth and explanation of reality.

Belief on the other hand is the power to know, Hence Belief carries with it the attributes of the believer, heightened and unraveled as the self purifies. Knowledge has a monotonic relation with belief and its attributes. Yet belief as fact cannot be measured in the form of cognition. This is true of both individual and collective beliefs, as of the community in the second case. On the other hand, the effect of belief through the attributes of believing can impact on the exercise of knowledge gained from degrees of participation, complementarities and paired linkages and interrelationships. All these lead to levels of knowledge formation. Subsequently, such knowledge-flows caused through a discursive milieu in the light of unity of knowledge as the episteme, induce relational entities.

1. Belief, Knowledge and Value

An example here is of an economic good. An economic good in the context of economic rationality can be any good that has price as value and gives satisfaction to the consumer and revenue to the seller. This is not a definition in any sense in the context of the moral and ethical embedding of the good. Rather, in this case an economic good is an embedded social, ethical and moral good whose use by consumers and producers results in simulation of wellbeing over streams of knowledge-flows, as the episteme of divine unity combines with the organization of markets for the 'good things of life'. Ethical consumer preferences, social menus and sustainability are realized through formal discourse, social action and response (Choudhury, 1994).

In this way, the two conditions that Ghazali attached to the determination of an event become functional. These are firstly, the systemic state of the issue and problem as observed, the extant of prevailing information received. Secondly, there is the institutional connection with the state of the observed entities. This is used to guide the transformation and improvement of the system along heightened paths of belief in the unity of the divine laws. Intrinsic observation as state of the system is thus combined with the institutional function to enact ways and means, policies and programs that affect necessary ethical and moral values. In the end, an economic good as a socially, morally and ethically embedded entity becomes the result of a guided market exchange mechanism in view of the objective of simulating wellbeing under conditions of knowledge-flows that are continuously generated by polity-market interaction (Choudhury, 2002) in the transformation process. An ethical good of this kind has a knowledge core. This is an ordinal knowledge value that can be simulated by institutional activity in concert with the state of markets as systems of social contracts (Choudhury, 1996). Another example is of capital as a 'good'. Then too such an ethicized good in an ethicizing resource formation venue becomes a spiritual capital (Zohar & Marshall, 2004).

Since multimarkets, resource mobilization and polity-market interaction, integration and evolution are forms of systems with learning in them in the quest for ethical and moral excellence; such systems become neurocybernetic in nature with learning in them. In this way, all forms of system-oriented issues and problems that rely on relational world-systems can be embedded in moral and ethical values. Machines as capital goods indeed have such social positioning as well (Choudhury, 1998).

Every such induction of an artifact and social relation by moral and ethical embedding carries with it a measure of knowledge-flows premised on the epistemology of unity of the divine laws; the consciousness of belief with its attributes that enable the observation and explanation; and the occurrence of an event as a socially embedded artifact. The combined result of all these is the

177

simulation of social wellbeing with the help of the events described as morally and ethically embedded entities that unify relationally.

A symbolism signifies this kind of relationship between Primal Knowledge, Belief (Consciousness), Knowledge-Flows, Event and Evolution across continuums of knowledge, time and space:

$$\Omega \rightarrow B(A) \rightarrow \theta(B(A)) \rightarrow X(\theta(B(A))) \rightarrow W(\theta(B(A)), X(\theta(B(A)))) \rightarrow \text{ simulation and}$$
$$\text{continuity}$$

In order to close this string, so as to give it the possibility for evolutionary equilibrium and thus problem solvability over interacting, integrating and evolutionary domains of knowledge and knowledge-induced artifacts, we write,

$$\Omega \rightarrow_s B(A) \rightarrow \theta(B(A)) \rightarrow X(\theta(B(A))) \rightarrow W(\theta(B(A)), X(\theta(B(A)))) \rightarrow \text{ simulation in}$$
$$\text{continuum} \rightarrow \Omega \quad (6.1)$$

Ω denotes the fundamental epistemology.

$\rightarrow_s$ denotes the transmission of Ω through the medium of the Sunnah (Prophetic guidance), blessed Guidance by the life, sayings and actions of the Prophet Muhammad, denoted by 's'.

B(A) denotes Belief premised on attributes that derive their qualities from Ω.

$\theta(B(A))$ denotes knowledge-flow premised on Ω (unity of divine knowledge) and actualized through 's' and B(A).

$X(\theta(B(A)))$ denotes the vector of state variables interacted with polity-market influence conveyed through $\theta(B(A))) \in \Omega$.

$(\theta(B(A)), X(\theta(B(A)))$ denotes an event (vector of events taken together as a point of polity-market interaction, integration and causing evolution (IIE-process as mentioned in earlier chapters) in learning and knowledge-induced things).

W(θ(B(A)),$\mathbf{X}$(θ(B(A))))) denotes wellbeing evaluation function for testing the degree of unification of knowledge in the system invoked by the issues and problems under investigation.

Hence W(.) measures the degree of unifying linkages caused by circular causation interrelations between the $\mathbf{X}$(θ(B(A))-vector variables. All such vector variables are embedded in knowledge-flows θ(B(A)), which in turn denotes the limiting value of discoursed knowledge-flows. Such knowledge-flows expressed in ordinal values in the institutions with formal processes are primarily premised on the epistemology of unity of knowledge.

The evolutionary property of the learning string is shown by the simulation in continuum.

$\rightarrow$ signifies causality from one part of the system to the next. In this way, 'simulation in continuum' signifies repetition of similar 'bits' in the chain (string). Thus circular causation is realized between phases of learning and between the knowledge-induced entity relations in the growing branches of the knowledge-tree.

The string, "$\Omega \rightarrow_s$ World-System$\{\theta, \mathbf{X}(\theta)\} \rightarrow \Omega$", explains the passage and sustainability of the world-system through knowledge of the divine laws across continuums and continuity of time..

The knowledge-induced 'bit' given by expression (6.1) is shown alternatively as process-oriented progression followed by repetition and reflexivity of similar processes across 'simulation in continuums'. The process continues until the closure of the string at Ω:

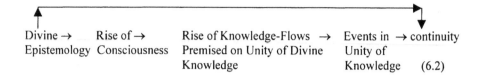

Divine →   Rise of →   Rise of Knowledge-Flows   →   Events in → continuity
Epistemology   Consciousness   Premised on Unity of Divine   Unity of
                                 Knowledge   Knowledge   (6.2)

Expression (6.2) is the manifestation given by Ghazali in terms of his understanding of the relationship between the divine laws, consciousness, and the path to knowledge, the purpose of existence and the knowledge-centered world-systems. All these he premised on the core of the Shari'ah. The peripheral part of the Shari'ah is humanly discoursed in the light of the divine core.

Details on the dynamics of relations and the formalism underlying (6.1) will be developed in Chapter 7. This kind of manifestation of systemic reality as explained by Ghazali in terms of his understanding of causality and reflexivity, as mentioned earlier, defines the knowledge-induced 'bit' of the entire complex strings and nexus of learning across interrelated systems and entities. Expression (6.1) then becomes a spanning system of expanding branches of the grand knowledge-tree (Choudhury. 2000).

Neurocybernetic Construction with Ghazali's Knowledge-Simulated 'BIT'
Neurocybernetic construction for explaining systemic reality and problem-solving within it are based on the indispensable properties of continuity and continuums of relations. We have explained this point before. Beyond continuity over time the basis of consciousness in systems thinking takes us into continuums, where time loses its independent role. This is the rationalist perception of the entire occidental epistemologies and systems treatment of complex problems. Continuums instead, are defined, structured and changed by the primacy of knowledge-flows and relations caused by them. This phase of the learning process is completed by pervasive unity of knowledge gained through pervasive participation, linkages

and complementarities. None of these is the property of either methodological individualism or Darwinian mutational forms.

Time exists in such systems as a medium to record change. Change is caused principally by knowledge and learning. Spatial activity in the time-driven system is caused by the automatic endogenous nature of the system, as of markets. The endogenous role of learning through institutions and causality remains absent. Thereby, institutions are either exogenous to the state of the system under investigation or they imitate the system. In so doing, as in the occidental economic and social orders, institutions repeat the methodological individualism and Darwinian mutation that are perceived to be in the state of the system under investigation. In other words, despite the essentially unified nature and scheme of things, the rationalist and neo-liberal view of alienation and individuation in such systems present a disparate understanding of systems. Besides, they monitor the same in keeping with the perception of alienation and competition in mind. Consequently, the policies and programs so emanating from the institutions structure govern change by an inverted perception of reality.

In the learning universe of unity of knowledge, time is a recorder of events. Time is endogenously generated in the system of knowledge-induced change. Now learning as the causation of change is also endogenous in the pervasive entity-system circular causation relations subsumed in unity of knowledge.

The concept of space now becomes different in the two cases of time, namely as exogenous and endogenous entities. Space is defined and change enacted in it exogenously by the time-variable. The question as to how change comes about, being a matter of explanation by a process understanding of system behavior, remains absent in the simply time-driven system.

Contrarily, in the case of the knowledge-driven system of pervasive relations induced by unity of knowledge, the perception of time is caused by the

181

pre-existence of knowledge. Hence time becomes endogenous in connection with knowledge through circular causation in such a system. Consequently, an observation and event are together caused by the simultaneous existence of knowledge, which is also simultaneously recorded by time. Time now equals knowledge-flow as a moment of recording the change, but not as the cause of change. This is the same as to proclaim, that at the moment of knowledge 'I' perceive the state of the world in time.

Formalism in Knowledge, Event and Time in Contrasting Universes

The following formalism explains the differentiated approaches to the treatment of knowledge, event and time, at a point in time:

1. *Occidental socio-scientific thought*

$\mathbf{x} = \{x_1(t), x_2(t), \ldots, x_n(t)\}$ is a state vector of variables; t is time variable.

$\theta = \theta(t)$, but reflexivity does not exist. Hence, $t = t(\theta)$ remains undefined. Consequently, $\mathbf{x}(\theta) = \{x_1(t(\theta)), x_2(t(\theta)), \ldots, x_n(t(\theta))\}$ remains undefined. Thereby, $d\mathbf{x}/d\theta = \Sigma_i(\partial x_i/\partial t).(dt/d\theta)$, remains undefined. Consequently, knowledge and space configurations lose meaning in terms of time as function of knowledge, at a moment of time.

2. *Explanation by the phenomenology of unity of knowledge*

Redefine, $\mathbf{x} = \{x_1(t), x_2(t), \ldots, x_n(t)\}$, with reflexivity between θ and t in the sense that θ primarily explains event and change. 't' records such states of change. Now, $d\mathbf{x}/d\theta = \Sigma_i(\partial x_i/\partial t).(dt/d\theta)$ is meaningful. This implies that, $d\mathbf{x}/dt = \Sigma_i(\partial x_i/\partial \theta).(d\theta/dt)$. Yet these two expressions are not equivalent. They convey meanings in the process orientation of systems that are different from the case 1 above. The interconnectedness between these two expressions is in the sense of circular causation in the universe of $\{\theta,t,x\}$. We explain this circular causation now.

$$\theta \to t(\theta) \to \quad x(t(\theta)) = x(\theta) \qquad\qquad (6.3)$$
$$\downarrow f_1 \qquad\qquad \uparrow f_2$$
$$\theta' = \theta'(t) \to x'(\theta'(t)) = x'(t)$$

Thus the reflexivity problem of knowledge-time reversibility means that there is a non-trivial functional relation $f_1(.)$ mapping $(t(\theta), x(t(\theta)))$ to $(\theta'(t), x'(\theta'(t)))$. Conversely, let f_2 denote a similar functional relation mapping $(\theta'(t), x'(\theta'(t)))$ to $(t(\theta), x(t(\theta)))$. The condition, $f_2 \circ f_1 \neq I$, the identity mapping, holds, because $t(\theta) \neq \theta(t)$. Consequently, $x(\theta) \neq x'(t)$. This result is contrary to the non-reflexive result of Case 1 mentioned above, wherein, $t(\theta)$ remains undefined.

Ghazali's Questions on Knowledge-Time Reflexivity

Ghazali's disagreement with the peripatetic philosophers in Greek lineage made him to argue that events occur in time because time is primarily ordained by God. That is, knowledge under the divine laws creates and re-originates time. Thereby, the world is explained in relation to such a created time by the impact of primal knowledge. We recall here that knowledge-flows are assigned institutionally as ordinal values through the medium of participation. Such participation can be subtle trans-empirical unification relationships between entities in the intrinsic state of nature. They are ontologically unraveled and assigned ordinal values through socio-scientific discourse keeping in sight the presence of state. Therefore, with the impact of primal knowledge in creating and originating time and the world in it, there also comes about the original impact of knowledge-flows premised on unity of knowledge on the created and originated entities.

These include time. We thereby have different views of time – psychological, transcendental, relativistic, quantum, market-period time, and clock-time.

Several categories of time were considered by Ghazali, extending thereby his understanding of the relationship between knowledge and time both in the primal and temporal sense of the created and re-originated worlds. Ghazali argued against the peripatetic philosophers of Greek lineage that the world was created in temporal time. Hence it was possible for the world to receive the primal impact of knowledge and its ordination of the world, including temporal time. But before the worlds were created and originated there existed abstraction as phenomenon (al-Ghayb). This must be included in the total equation of cosmic reality. Yet that nature of time was non-temporal. Hence in mathematical logic of continuum and continuity the abstract order of time was topological in nature.

Knowledge ensuing from the divine will caused relational orders to develop through mappings between the trans-empirical realities (abstraction) of the sublime worlds before the creation of the temporal worlds. In such trans-empirical realities there was still the existence and continuity across continuums of relational domains. In such universes knowledge of the participating, and thus unifying world-systems existed. But such relations appear as mathematical formalism. They explain fundamental truth of causality in the origination of trans-empirical realities leading to empirical quantification.

Likewise, by the end of all world-systems nearing the completion of Ω in expression (6.1), a similar phenomenon with respect to the relation between knowledge, time and event exists as in the primal case of Ω as the fundamental epistemology of being. This Ω as the Event of the Hereafter marks the actualization of the super-cardinal essence of the primal Ω also existing in the realm of completion of knowledge and truth.

This reflexive nature of forward and backward relations necessitates the reverse mapping of such limiting universes of events in the world-systems. The result is the two-way causality of the very large-scale universal relations and circular causations that pervasively fill all possible nexus made up of branches of the knowledge-tree. This is equivalent to the spanning of the learning universes by the knowledge-induced 'bit' shown in expression (6.1). This two-way causality is shown by the forward causation ($\Omega \to$ World-System). This is combined by the end reflexive causation, (World-System $\leftarrow \Omega$). Combining these with the complete learning course, ($\Omega \to$ World-System $\to \Omega$), yields the reflexive result, ($\Omega \leftrightarrow$ World-System $\leftrightarrow \Omega$).

Nonetheless, there exists the interpretation of this chain from the two sides -- of the Primal Ω to the Hereafter Ω in the context of explaining learning and induction as process. Next, the equivalence of the Hereafter Ω with the Primal Ω exists as super-cardinal and open topological closure of the universes. No process can exist at these terminal points. Only in such topological super-cardinal and open events, time of the transcendental nature merges with primal knowledge. This is why, in an inspired saying of the Prophet Muhammad, we find the proclamation, and "I (God) am Time".

The formal derivation of the relationship between consciousness, knowledge, time and event in Ghazali's epistemology leads to a straightforward understanding of his criticism of the rationalist Muslim scholastics (Marmura, p. 29): "For is the circular motion, which is the foundation [of all temporal events], temporally originated or eternal?" The answer to this question as derived from the foregoing arguments is that the supreme and super-cardinal domain of Ω is process-less. It is the bundle of the divine laws that are explained and externalized by the Prophetic Sunnah (Guidance), which in expression (6.1) is denoted by 's'. In the process-less essence of Ω, knowledge and time are the same in their topological sense of the ultimate capacity to create and originate, not to be created or originated. Thereby, Ω of the Hereafter is the accumulation of *all* knowledge-flows of *all* the worlds, unraveled and hidden, into the completeness of Ω in the primal. The hidden worlds comprise the realm of God and His preserve of knowledge as well as the hidden world-systems of the rendered world-systems. The super-cardinal topologies of Ω in both the primal and penultimate sense form the fundamental epistemology that remain exogenous but are recalled to initiate all emergent processes. The two Ω's are equivalent to each other in the topological and super-cardinal open world-systems induced by knowledge-flows arising from Ω and enabled by the transmission of 's'. The circular motions as shown in expression (6.3) denote the ordained consequences of the exogenous action of the divine will Ω.

Cybernetic Consequences of Knowledge-Time Reflexivity Relationship

System and cybernetic theory deals with Realtime conception and problem-solving. In Turing Machine (Turing, 1936), Godel's self-Referencing model of printable sentences (Smullyan, 1992), and the supercomputer algorithmic logic (Kuskov, 1999), the Realtime concept and application is of a pervasive nature, enabling complex artificial intelligence designs to form and be enabled for problem-solving. In reference to the above explanation, we define Realtime as the order of continuous circular causation explained in expression (6.3) that interrelate knowledge and time reversibly across learning domains in unity of knowledge.

Example of a Pursuit Game in Clock Time: Discrete Case

In a situation of learning and time-recording, one can envision a game being played across indefinitely large number of mazes (nexus) of interrelated events. In

186

the hide and seek pursuit across such mazes a search for the thief can take two forms.

Firstly, consider a clock-time game being played out by the pursuer pursuing the thief. In slow motion, this would look like a graduated movement of the pursuer pursing only up to a point of nearness of apprehending the pursued, when suddenly the thief escapes in another alley of the maze. As this experience continues, a list of differential times and differential distances is recorded between the pursuer and the thief. The thief is never caught and the pursuit never ends.

The result then is that there is a concern on the measure of the distance between the pursuer and the thief. The system starts to wonder how much differential time it will take to apprehend the thief (or likewise the time elapsed up to that point of the pursuit). In this infinitely played game we note that an event (pursuit) in space is of the essence, and time measures the attained state of that event in space. Consequently, the differential (Δ) distance between the pursuer and the thief, say x, is determined by the differential (or elapsed time up to the point of event). That is, $\Delta x = \Delta x(t)$ (or replace t by Δt).

What is the knowledge implication in this infinitely played game? The distance between the pursuer and the thief that evades apprehension of the latter (termination of the game) is defined as having an inverse relationship to the knowledge difference. That is, if θ^* is the knowledge required to terminate the game, and if θ is the moving (i.e. simulated) knowledge-flow, then $\Delta\theta = (\theta - \theta^*)$, with $\Delta x \propto 1/\Delta\theta$ ($\propto$ means proportionate to). Likewise in terms of time as given above, since $\Delta x \propto \Delta t$, therefore, $\Delta t \propto 1/\Delta\theta$. θ^* is a knowledge-flow that would have apprehended the thief and ended the game. In an infinitely played game ($\theta \rightarrow \theta^*$) continuously without $\theta = \theta^*$ in any particular reconstructed strategy.

On collecting these relations we obtain say, $\Delta t = A.(\Delta\theta)^\alpha$, and $\Delta x = B.(\Delta\theta)^\beta$. By combining the foregoing result we obtain $\Delta x \propto \Delta t \propto 1/\Delta\theta$. Here, A and B are suitable estimable parameters; $\alpha < 0$, $\beta < 0$ are negative coefficients in the general case of complex pursuits in the infinitely played game. In the special case of proportionate distancing, $\alpha = \beta = -1$. x and t determinations (simulation) are thereby functions of the prior determination of θ.

We ask the question: Can we describe the above game in terms of θ as functions of t and x? This is possible, for an attained state of (x,t) will bring about strategy change invoking new knowledge formation in order to attain a new state, say (x',t'). Consequently, by explaining in a way similar to the above case, the following game is a continuity of the former one along phases of learning. That is, $\Delta\theta' = A.(\Delta t')^{\alpha}$, and $\Delta\theta = B.(\Delta x')^{\beta}$.

Conclusively, circular causation exists:

$$\Delta t = A.(\Delta\theta)^{\alpha}, \text{ and } \Delta x = B.(\Delta\theta)^{\beta}$$
$$\Delta\theta' = A.(\Delta t')^{\alpha}, \text{ and } \Delta\theta = B.(\Delta x')^{\beta}. \tag{6.4}$$

Thereby the implications of expression (6.3) hold.

Note that every phase of the pursuit in the infinitely played game is a learning process over the previous one, and hence a point of re-strategizing. Therefore, each knowledge-flow gained must be higher than the previous ones. Transaction cost in the pursuit system (leakages) will be reduced thereby. Only in this way there can exist evolutionary equilibriums for explaining the game in terms of knowledge-induced state variables.

Two conditions of the circular causation (reflexivity) relations are to be noted as essential for the existence of expression (6.4) over continuums of pursuits as events in order to realize evolutionary learning equilibriums.

According to the essence of consciousness in his system orientation, Ghazali argues that the state of an event and its re-strategizing are ordained by the divine laws. They are explicated by the Sunnah to bestow behavior and observability. The learned worldly discourses premised on unity of knowledge interpret the epistemological inquiry. Consequently, in reference to Ghazali's episteme, the first condition of the nature of the pursuit game with evolutionary equilibriums in it is a selection of strategies, which are to be premised on the fundamental texts of the episteme. These texts are the Qur'an and the Sunnah in this order of primacy. The rules so discovered from the texts are then rendered to learned discourse.

In the pursuit game mentioned above the strategies for discourse comprise discovering ways and means of demoralizing falsehood and evil by putting fear and terror (Truth, Goodness) in the heart and mind of the thief (Falsehood, Evil). Such selections are ratified by the crowning Qur'anic principle of struggle in the path of truth with might and main and the divine promise of victory of truth and goodness over falsehood and evil. The meaning of truth, goodness, falsehood and evil explained in the Qur'an in precise terms.

The second condition is that continuity in the circular causation between knowledge-flows and the knowledge-induced time and state variables must prevail across continuums of the pursuit events. The continuum is defined here by the continuously recursive circular causation interrelations between the entities of the variables in the vector, $\{\theta, t(\theta), x(\theta)\}$, in view of the reflexivity shown in expression (6.4) and earlier in expression (6.3).

Example of a Pursuit Game in Realtime: Continuous Case

As defined earlier, Realtime signifies continuous evolution of circular causation *inter*relationships between reflexive entities including knowledge and time. These interact, integrate and evolve by circular causation across knowledge, time and space continuums. When the above-mentioned pursuit game is cast in Realtime, the circular causation interrelations exist simultaneously with pervasive

189

complementarities between the entities and their relations. In such a case, the requirements of an efficient game with evolutionary learning equilibriums are readily satisfied. Leakages (transaction cost) are reduced in the system. Learning is heightened in the game. Thereby, learning equilibriums evolve smoothly and speedily. Diversification and expansions by the interrelating sub-systems act as automatic controls for self-organizing the systemic relations and maintaining the evolutionary learning character of the system in the framework of unity of relations. Overarching self-organizing knowledge-flows and entity relations sustain the momentum generated by the episteme of unity of knowledge in the gaming system. These properties bestow logical formalism to the system in the midst of growing consciousness resulting in a continuous production of knowledge-flows. Problem-solving is thus enabled and explanatory power of both empirical and trans-empirical interpretations is increased.

The same arguments and explanation given in the last section hold up in the case of continuous games. Let us say that the present game is played out as a cat-and-mouse game in virtual reality. Except that in this case we must re-write expression (6.4) and thus interpret it in terms of continuously interdependent and recursive values of $\{\theta, t, x\}$ as follows:

$$dt/d\theta = A.\alpha.(\theta-\theta^*)^{\alpha-1}, \text{ and } dx/d\theta = B.\beta.(\theta-\theta^*)^{\beta-1}$$
$$d\theta/dt = A.\alpha.(t-t^*)^{\alpha}, \text{ and } d\theta/dx = B.\beta.(x-x^*)^{\beta}. \tag{6.5}$$

When x is a vector of variables then its Jacobian with respect to θ and t must be non-vanishing and must exist in the mathematical sense (Hogg and Craig, 1995). Besides, the expression (6.5) will become much more complex if the coefficients are also perturbed by θ-values. They are then observed as Brownian Motion over time and space.

Furthermore, in the case of the knowledge, time and space continuums expression (6.5) will yield,

$$\theta = \alpha A . \int_\theta^{\theta^*} \int_t^{t^*} (t(\theta) - t^*(\theta^*))^\alpha dt \, d\theta \qquad (6.6)$$

$$t = \alpha A . \int_\theta^{\theta^*} \int_t^{t^*} (\theta(t) - \theta^*(t^*))^{\alpha-1} dt \, d\theta \qquad (6.7)$$

$$x = \beta B . \int_t^{t^*} \int_\theta^{0^*} (\theta(t) - \theta^*(t))^{\beta-\alpha} dt \, d\theta \qquad (6.8)$$

For continuous infinite games we would have, $\theta(t) < \theta^*(t^*)$ for $t < t^*$. However, every given $\{\theta^*, t^*, x^*\}$-value belongs to a specific learning phase in the interactive, integrative and evolutionary (IIE-process) 'bit'. Consequently, (6.6)-(6.8) defines evolutionary approximation for $\{\theta, t, x\}$ in the phase that is learning in the 'bit'. To evaluate (6.6)-(6.8) over the entire nexus of $\{\theta, t, x\}$ we need to evaluate the social wellbeing function in which the variables are introduced. The social wellbeing function is then simulated over interaction (θ), integration (θ^*) and evolution (new θ-value). The same is true of (t,x) in terms of θ-values.

Ghazali's Neurocybernetic Implications from the Learning Games

Ghazali explains consciousness as the determining originality of mind on which all else is premised. The above examples are simply cases of problem-solving in reference to the worldview of unity of divine knowledge. In regards to the general case of consciousness underlying insight as experience in unraveling deeper truth gained by the exercise of knowledge in Tawhid, Ghazali writes (Karim, undated, p. 237-38 Vol. IV) (edited): "The first stage (of Tawhid) is like outer cover of a cocoanut, the second stage is the inner cover of a cocoanut, the third stage is the kernel of a cocoanut and the fourth stage is the oil of the kernel. The first stage of Tawhid is to utter by tongue, 'There is no deity but God.' The second stage is to

191

confirm it by heart. The third stage is like kernel, which can be seen by inner light (Kashf). This is the stage of those who are near God. The fourth stage is like oil in kernel. He sees nothing but God. This is the stage of the truthful. It is called Fana-fi-Tawhid or to lose oneself in Tawhid."

Another Mundane Example of Spiritual Consciousness in Problem-Solving

In the problem of cooperative game played, say between two innocent persons A and B, neither of them ought to confess to a charge of crime on them. Therefore, the two will collude, and their firm stand on innocence will drive them to consider a finite game to end the judgment against them. If this was not to be true, A and B would cheat each other by trying to implicate the other in the crime by silent deception. The two would then play a two-person cooperative game of the Prisoners' Dilemma (Lutz & Lux, 1988). The game would then end in a cooperative situation in which both prisoners confess not to have committed the crime. A breach of ethics results in such a confession. In the ethical case the innocent would not be in a contract with the real convict for his reprieve. Thus a Nash Equilibrium in the case of Prisoners' Dilemma is not a viable solution in the ethical case. The way out of this impossibility is for a change in social contract that would reprieve the convicts if they made honest confessions. The loss from such confession would not be in excess of that in the case of the cooperative alternative of Prisoners' Dilemma game.

The ethical prisoners' contract are now of two types in finite game. Firstly, when both A and B are innocent, they do not have to confess to any charge. A new contract is needed to ease their burden by discourse, in which case the view of the game-table from the side of the prison warden gets A and B out of prison by their cooperative confession truly denying the charge. But this is the result of mutual consultation between the prisoners. It is a condition contrary to the nature of cooperative game played out in the problem of Prisoners' Dilemma.

Contrarily, if consultation between A and B is allowed for, their individual payoffs are embedded in ethical confirmation, which forms the knowledge embedding of the payoff. Thus, it is the primacy of the interactive and consensual confirmation based on the ethical embedding of the payoff that solves the game in finite time. That is, at once as the consensus is attained through the prisoners' consultative process, no further evolution in learning is required, as no re-contracting is needed. The evolutionary game under learning by interactive and integrative relations ends.

In the case of the problem of Prisoners' Dilemma, the ethical question is rejected by cheating and elusive behavior in the midst of independence of discourse between the prisoners. The ethical case of truthful confession does not abide in determining the Nash Equilibrium of the Prisoners' Dilemma game. Consequently, the moral and ethical induction of contract in this game necessitates guarantee of perks to the truthful prisoners in the amount of loss that does not exceed the Nash Equilibrium cost as in the disparately cooperative situation of lying to escape. A finite game is thereby enacted by a change in the administration of the costs and payoffs to the prisoners. This is a kind of relational conduct between the prison authority and the prisoners that interrupts the condition of social independence between the prisoners and the prison authority, which otherwise exists in the Prisoners' Dilemma game. Acting upon and responding to a case of social contract accomplishes the matter of growing consciousness in the system. It is generated by interaction, integration and social dynamics between the prisoners and the prison authority.

Can Consciousness be Possible in the Real World?

In the domain of moral consciousness, the errors of worldly actions, beliefs and egos are made manifest for moral rectification. Social order and a good society require balance between truth, self and the other in order to form and reform. All good institutions that promote common wellbeing and human sustainability aim at enacting such policies and programs that raise moral and ethical consciousness in the midst of material balance. Then the material artifacts become spiritually embedded by moral consciousness, as was explained earlier in terms of individual and social preferences on choices involving markets, exchangeable, wealth and capital formation.

In the Prisoners' Dilemma game in which at least one of the prisoners cheats by treating the act of lying as socially acceptable when it is not, is reflected in Ghazali's characterization of worldly errors caused by such deception of self against reality. In this regard (Karim, p. 348-49, Vol. III) writes (edited): "The thing which brings peace (exoneration from prison by the liar) of mind consistent with low desire (lying) is error. Nature (Prisoners' Dilemma and Nash Equilibrium in this case) is inclined to its entertaining doubt and falling in devil's snares." In the project of computing reality all such morally and socially unacceptable methods and methodologies must be critically questioned, then rejected and replaced by meaningful ones.

In the area of game theory, moral reconstruction of methods and methodology can be the study of knowledge, evolutionary equilibriums and games (Osborne and Rubinstein, 1994, chapter 5). Moral consciousness endogenizes knowledge by causing the participating actors, natural entities and contracts (states and policy variables) to be part of the information input-output system with recursive feedback existing and evolving in the system. This is the nature of endogenizing moral and ethical conduct in the social contract, as in the case of the participating rule-formation between prisoners and the prison authority, so as to exact truth and induce consciousness, while giving perks for inciting the moral conduct. Contrary to this is the state of relational independence that excites the possibility for cheating and error of consciousness in the lying prisoners.

Indeed, in the project of Computing Reality the conscious character inducing methods and methodology must be maintained. Choudhury (1999) points out some of those methods that can be applied due to erroneous methodology for solving social problems. But methodology is of primal importance on the question of moral, ethical and social issues. If so understood, then the relevant methodology will give rise to methods and rules that are mutually compliant. The distinction between the methodologies and methods of the prisoners' game theory in the two contrasting cases is the case in point.

Consciousness of the Empirical and Trans-Empirical Worlds

A distinguished lecture by Belal Baaquie (2005) treated the dialectics of the Qur'anic paired universes – paired between the good things that are compliant with the divine and moral law, and between the bad things by the same criterion. His example is of the Black Hole, whose inner space cannot be studied by physical observations. But mathematical String Theory can extract the relational dynamics of such a 'black' interior where the laws of physics become singular.

Causality involves multiple interactions. In the concept of Tawhidi unity of knowledge as championed by Ghazali, we have identified that the underlying methodology is governed by the learning (process) properties of interaction, integration and evolution (IIE). In such a case, the dialectic of the paired universe establishes itself by circular causation between the entities and their relations, as they learn and evolve over continuums. The problem of the Black Hole in respect to the method of circular causation in the framework of epistemology of unity of divine knowledge arises from its very definition.

Stephen Hawking (1988) explains the mathematical concept of Black Hole in this way: In certain regions of the cosmos very strong gravitational fields exist, such that the light in its proximity existing either as particles (photons) or waves are drawn into the massive gravitational fields. It is assumed here that gravitational force exceeds the finite speed of light either as particles (Newtonian) or waves (Quantum Physics). Thus the largest star can be drawn into the black hole causing the death of such stars and indefinitely increasing the intensity and mass of the powerful and lumpy gravitational field. These are the black holes by virtue of the fact that "if light cannot escape (from them), neither can anything else; everything is dragged back by the gravitational field. So, one has a set of events, a region of space-time, from which it is not possible to escape to reach a distant observer. This region is what we now call a black hole. Its boundary is called the event horizon and it coincides with the paths of light rays that just fail to escape from the black hole." (Hawking, op cit p. 87).

In the case of circular causation existing as dialectic of reflexive relations, say between events A and B, occurring within the Black Hole, these must be identified. They may not be empirical facts, but can be mathematically studied as trans-empirical facts.

One such mathematical approach is by the exercise of consciousness. The methodology responding to this kind of examination and characterization of events within the Black Hole is Superstring Theory, the study of emitted responses from the Black Hole in the form of Quantum waves. The problem though is this. How does an observer identify not the physical, rather the mathematical entities in the string version of the Black Hole to establish reflexive relationship of paired and circular causality within it and then extract this information by a mathematical string? Since nothing that disappears into the Black Hole ever responds to the surface, therefore, either the string version of the study of causality inside the Black Hole is unreal and meaningless or it is simply subjective. In the end, the consciousness of the Black Hole remains hidden inside it and is not unraveled to the observer to serve a meaningful objective.

Consciousness of the kind that is simply subjectively hidden without even a flick of its unraveling is not within the domain of Ghazali's epistemology of oneness of the divine laws for explaining the universe. Indeed, such an unraveling is never perfect and complete except with God alone. But the unraveling of that absolute power is within the worldly equation and is attained in degrees. In this regard the Qur'an declares (43:51): "It is not given to any human being that Allah should speak to him unless (it be) by Revelation, or from behind a veil, or (that) He sends a Messenger to reveal what He wills by His Leave. Verily, He is Most High, Most Wise."

Therefore, in order to make sense of the dialectical reading of the interior of the Black Hole, the definition of the Black Hole must be changed so as to accommodate the possibility of externalizing the inner dynamics of circular causation inside the Black Hole generating information. The superstring theory as a method of such extraction by means of wave-theory of light and gravitation would then make methodological sense. Let us see whether a new definition is possible for the Black Hole in the light of this methodological necessity?

In his University of Dublin Lecture, Hawking (see Baez, 2004) acceded, that information can be preserved in the Black Hole, if the Black Hole is understood in a new way. The first kind is the trivial black hole. It evaporates and decays, thus releasing the information in it. The second kind is the non-trivial black holes, which remains permanent. Thus information decays inside it although it can be preserved. But the net result is that only the non-trivial black hole can be explained in terms of causality generated by information exchange between learning entities. Hawking (quoted in Baez, 2004) writes: "Information is lost in topologically nontrivial metrics, like the eternal black hole. On the other hand, information is preserved in topologically trivial metrics. The confusion and paradox arose because people thought classically, in terms of a single topology for spacetime."

We have shown by means of expression (6.6) that by the entire nexus generated by phases of recursive relations in the learning process of expression (6.1), and thereby (6.3), the entire learning processes is integrated by contour integration over the interactive, integrative and evolutionary nexus of topologies defining $\{\theta, t$ and $x\}$, Such topological relations are recursively formed and continued over continuums.

Thus, by opening up the closure of the Black Hole and making it interactive with the environment, just as for any other learning entity under condition of relational unity, the true meaning of consciousness is realized. Consciousness is then defined by expression (6.1) by relating this to the processes of the paired entities under study. Examples are of the paired relations between matter and mind, gravitation and particle/wave, energy and matter inside the Black Hole. All entities that fall within the Black Hole are energized but not to the point of their relational oblivion. Such entities and their relations have history of the past and future. They are events recognizable at the present, though partially. They have consciousness that unravels objectivity to the observer, who is within this system of information creation.

In the light of expression (6.1), now applied to the case of cosmic consciousness, the mathematical phenomenon of Black Hole can be defined as a mathematical entity that agrees with the relational dynamics of unity of knowledge over continuums of learning. The latter condition enables increasing quanta of information to be extracted, stored and used for the study of the Black Hole in relationship to its objective environment of multidimensional embedding.

Overarching Objectivity of Socio-Scientific Consciousness

In the overarching case, the social and human environment and the heart and soul are not left out independently of each other. Kant's insightful expression can be recalled on this issue of combining the cosmic with the mundane in the light of the moral imperative. Kant wrote (quoted in 1949, 261): "Two things fill the mind with ever new and increasing awe and admiration the more frequently and continuously reflection is occupied with them; the starred heaven above me and the moral law within me. I ought not to seek either outside my field of vision, as though they were either shrouded in obscurity or were visionary. I see them confronting me and link them immediately with the consciousness of my existence."

Kant's mind reflected Ghazali's epistemology. Ghazali explained that if an object under study is to be morally understood, it is necessary firstly to understand the meaning of its moral worth, then the entity under investigation. In this regard, if we are to study the Black Hole as having a moral consciousness in it, then we must understand the meaning and scope of moral consciousness in extant over the socio-scientific field of human inquiry.

In the project of Computing Reality, such consciousness is explained by the expression (6.1) and its extensions across continuums (nexus). Simultaneously, along with the understanding of the moral essence must proceed on the construction followed by evidential inquiry. These are both empirical and trans-empirical in nature comprising $\{\theta, t, x\}$. In this way, the embedding of the event and entity in moral value transforms them into substantive spiritual cosmic forms and induces the learning process premised on unity of knowledge into socio-scientific understanding. In term of Ghazali's translation we note the words (Karim, p. 25, Vol. I), "The proof of reason for the excellence of knowledge is this. If the word excellence is not understood, it is not possible to know the excellence of other things. For instance, if one desires to know whether Zaid is a wise man, he should know first the meaning of the word wisdom and then of Zaid or else he will go astray."

The implications of this example for the project of Computing Reality derived from a criticism of the theory of Black Hole are that mathematics is neither a methodology nor a method for the formation of consciousness. All such cognitions must be superseded firstly by the primal reality of God. This alone forms consciousness in the first place. Consciousness is then rendered to deepening possibility. The belief in God followed by growing consciousness together form the project of unity of knowledge as the universal paradigm of socio-scientific reasoning for understanding, studying and analyzing all issues and questions of existence. This kind of an intellectual venture forms a general system study. In it and by its methodology the partial case forms a general evolutionary equilibrium of sub-systems that continuously learn and span out into broader continuums (nexus) encompassing the environing reality. The final implication of such a challenging worldview is that the theory of the cosmos is not independent of the theory of human systems and other ones, nor conversely. They are together driven by the same law of unity of divine knowledge. This law provides the methodology. The resulting methodology then calls for adoption of the pertinent methods. This uniqueness is invariant by systems despite the diverse socio-scientific fields with their variant issues and problems.

CONCLUSION

In this chapter we have undertaken a study of one of the greatest thinkers on moral consciousness in relation to the divine laws, Abu Hamid [Imam] Ghazali. His works in the Twelfth Century challenged all of received Muslim thinking of the time that relied on Greek peripatetic ideas. Subsequently, during the Eighteenth Century European Enlightenment Ghazali's scholarly works became the groundwork for emulation by the scholastics. Ghazali resounded in Immanuel Kant's works particularly.

Ghazali as a chief exponent of the religious epistemological thought in Islamic scholasticism centered his idea of the relationship between God, man and the universe on human consciousness. This he explained in the most subtle, emphatic and beautiful imagery resting on the unity of divine knowledge, will and law (Tawhid). Man's quest for this unity of knowledge in self and in relation with others, and in identifying the core meaning of existence erected of divine origins of consciousness was the pursuit that thundered across Ghazali's epistemology. Thus Ghazali was an early cybernetic thinker who centered the project of studying the relational worldview between all things to be premised on Godly consciousness.

Today the moral meaning of all social actions and response that can make socio-scientific inquiry meaningful for humanity at large is argued out in this chapter to rest on revoking the moral consciousness and finding its substantive understanding in the episteme of unity of knowledge. This is the groundwork of neurocybernetic and system theory in the project of Computing Reality. This kind of moral and social context of substantive socio-scientific inquiry makes the study of Computing Reality a significant scientific research inquiry.

REFERENCES

Ali, Yusuf A. 1946. *The Holy Qur'an, Text, Translation and Commentary*, New York: McGregor & Werner.

Baaquie, B. Dec. 2005. "The Empirical and Trans-Empirical in Physics", distinguished lecture, International Conference on A Universal Paradigm of Socio-Scientific Reasoning, Asian University of Bangladesh.

Baez, J. quoting Hawking, S. July 25, 2004. "This Week's Finds in Mathematical Physics (Week 207)", *http://math.ucr.edu/home/baez/week207.html*

Bohr, N. 1985. "Discussions with Einstein on epistemological issues", in H. Folse, *The Philosophy of Niels Bohr: The Framework of Complementarity*, Amsterdam, The Netherlands: North Holland Physics Publishing.

Buchman, D. 1998. *Al-Ghazali, The Niche of Lights*, Provo, Utah: Brigham University Press.

Cantor, G. trans. P.E.J. Jourdain, 1955. *Contributions to the Founding of the Theory of Transfinite Numbers*, New York: Dover.

Choudhury, M.A. 1994. *Economic Theory and Social Institutions: A Critique with Special Reference to Canada*. Lanham, MD: University Press of America.

Choudhury, M.A. 1998. "A Socio-Scientific Theory of Continuous Machines", *Cybernetica* Vol. XL!, Nos.2/3/4.

Choudhury, M.A. 1996. "Markets as a System of Social Contracts", *International Journal of Social Economics*, 23:1.

Choudhury, M.A. 1999. "Methodological Conclusion", in *Comparative Economic Theory, Occidental and Islamic Perspectives*, Norwell, MA: Kluwer Academic, pp. 337-356.

Choudhury, M.A. 2000. *The Islamic Worldview, Socio-Scientific Perspectives*. London, Eng: Kegan Paul International.

Choudhury, M.A. 2002. *The Islamic World-System: A Study in Polity-Market Interaction*, London, Eng: RoutledgeCurzon.

Hawking, S.W. 1988. *A Brief History of Time, From the Big Bang to Black Holes*, New York: Bantam Books, Inc.

Hogg, R.V. and Craig, A.T. 1995. "Extensions of the Change-of-Variable Technique", in *Introduction to Mathematical Statistics*, New Jersey: Englewood Cliffs, pp. 186-192.
Kant, I. ed. C.J. Friedrich, 1949. "Critique of pure reason"; also "Critique of judgment", in *The Philosophy of Kant*; "Reason within the limits of reason", in *The Philosophy of Kant*; "Idea for a universal history with cosmopolitan content", in *The Philosophy of Kant*, in C.J. Friedrich Ed. *The Philosophy of Kant*. New York, NY: Modern Library.

Karim, M.F.. Undated. *Ihya Ulum-Id-Din* in. Lahore, Pakistan: Sh. Muhammad Ashraf.

Kuskov, A. 1999. "From the Anthropic Principle to the Global Strategy", *Kybernetes: The International Journal of Systems and Cybernetics*, 28:6/7, pp. 753-762.

Lutz, M. & Lux, K. 1988. "The Problem of Self-Interest and the Humanistic Response", in *Humanistic Economics, the New Challenge*, New York, NY: The Bookstrap Press, pp. 64-88.

Marmura, M.E. 1997. *Al-Ghazali, the Incoherence of the Philosophers*, Provo, Utah: Brigham University Press.

Osborne, M.J. and Rubinstein, A. 1994. "Knowledge and Equilibrium", in *A Course in Game Theory*, Cambridge, MA: The MIT Press, pp. 67-86.

Qadri, C.A. 1988. *Philosophy and Science in the Islamic World*, London, England: Routledge.

Russell, B. undated. "The Philosophy of the Continuum", in *Principles of Mathematics*, New York, NY: W.W. Norton, pp. 346-354.

Smullyan, R.M. 1992. *Godel's Incompleteness Theorems*, New York: Oxford University Press.

Turing, A.M. 1936. "On Computable Numbers, with an Application to the Entacheidungsproblem", *Proceedings of the London Mathematical Society*, Ser. 2, Vol. 42, pp. 230-265.

Zohar, D. & Marshall, I. 2004. *Spiritual Capital*, San Francisco, CA: Berrett-Koehler Publishers, Inc.

TECHNICAL APPENDIX TO CHAPTER 6*:
THE TOPOLOGICAL SUPERSPACE OF Ω

Let Ω denote the topological superspace of the stock of knowledge, having the property that it includes both its subspaces denoted by $\{\theta\}$ and their complements denoted by $\{\theta'\}$. θ denotes knowledge-flow; θ denotes de-knowledge (mathematical complement as opposite of θ). Thereby,

$$\{\theta\}\cap\{\theta'\} = \phi. \text{ Thereby, } \{\theta\}\cup\{\theta'\} \neq \phi. \tag{A6.1}$$

These conditions apply most generally to the totality of agents, systems, variables and their relations (not shown).

Let $\{\mu\}$ denote a measure-theoretic and order preserving function defined on $\{\theta\}\cup\{\theta'\}$. It has the properties,

$$\mu[\{\theta\}\cup\{\theta'\}] = \mu\{\theta\}\cup\mu\{\theta'\} = \{\mu(\theta)\}\cup\{\mu(\theta'))\} \neq \phi. \tag{A6.2}$$
$$\text{Also, } \mu[\{\theta\}\cap\{\theta'\}] = \mu[\{\theta\cap\theta'\}] = \phi. \tag{A6.3}$$

These expressions bring out the property of perfect independence (disjointness) between the two subspaces, which nonetheless are subspaces of Ω. Because Ω is a topology, all its subspaces are topologies and $\phi \in \Omega$.

When truth and falsehood are not well-determined, as in the case of indeterminate deliberations on issues through discourse, then it is possible that,

$$\{\theta\}\cap\{\theta'\} \neq \phi; \{\theta\}\cup\{\theta'\} \neq \phi; \tag{A6.4}$$
$$\mu[\{\theta\}\cap\{\theta'\}] \neq \mu[\{\theta\cap\theta'\}] \neq \phi \tag{A6.5}$$
$$\mu[\{\theta\}\cup\{\theta'\}] = \mu\{\theta\}\cup\mu\{\theta'\} = \{\mu(\theta)\}\cup\{\mu(\theta'))\} \neq \phi \tag{A6.6}$$

The conditions (A6.4)-(A6.6) will establish themselves in the case of either $\{\theta'\}\rightarrow$ falsehood, as the discourse continues over interactions in the learning process. Otherwise, $\{\theta'\}\rightarrow \{\theta\}$, in the same cases of interactions taking place over discourses. The above transformations under the measurable mappings $\{\mu(.)\}$ comprise the relationships of complex world-systems that are interrelated in the sense of $\{\theta\}$ and $\{\theta'\}$ taken separately, or when $\{\theta'\}$ converges either to a well determination between $\{\theta\}$ and $\{\theta'\}$, as interactions proceed across discourses, as the case may be.

---

*Choudhury, M.A. 1999. "A philosophico-mathematical theorem on unity of knowledge", *Kybernetes: The International Journal of Systems and Cybernetics*, 28:6/7, pp. 763-776.

CHAPTER 7: WORLDVIEW AS UNIVERSAL PARADIGM IN UNITY OF KNOWLEDGE

In this chapter we will investigate the foundations, formalism and functional application of what we understand by the worldview as the universal paradigm, in contrast to the ideas of normal science, paradigm and scientific revolution (Kuhn, 1970). Through a critical examination of various ideas along these lines we will come to conclude that the universal paradigm is deeply and invincibly entrenched in the epistemology of unity of knowledge founded in the divine laws. We will argue that in all of religo-scientific history it is only in the Qur'an that the pristine understanding and rendering of an epistemological approach to the precept of the worldview as the universal paradigm is laid down in the most detailed way. We will study this premise in this chapter. In the end, this chapter will bring out the phenomenological model of unity of the divine laws in relation to every world-system. The conceptual and functional nature of the worldview methodology of unity of knowledge is the sound basis for the epistemology of Computing Reality.

TOWARDS UNDERSTANDING THE PRECEPT OF THE WORLDVIEW

The ways toward coming abreast with the worldview, now defined as the universal and absolutely irreducible paradigm of all paradigms are two. Firstly, it is to note deeply the indispensable validity of the Oneness of God in the universal scene. This will embrace both the individual conscience and the public order of all the sciences. This confirmation arises from a re-emanation of private and public activity in the scholarly, political and community fronts. The second way is as Kuhn has explained how scientific revolution is established. It is to carry on vigorous activities in the project of the worldview by committed members, students, scholarly groups and the scientific forums ventilating that worldview.

The important elements of such a propagation of the worldview will exist at three levels. Firstly, the epistemological level is the primal. This will induce renewed awareness and consciousness in the beholder. In our project this consciousness is of beholding, understanding and applying the foundation of unity of divine knowledge to world-system studies. Thus the meaning of the divine functioning in such a foundational knowledge is of the essence in accepting the divine roots of knowledge. Without this, in connection with the phases of knowledge development that follows, recourse to divine knowledge for world reconstruction is not possible. Divine knowledge must therefore be meaningful in terms of its benefits for the fullest comprehension of reality and useful in developing inferences from it for life, existence and experience.

Secondly, the ontology emanating from the epistemological phase of thought as the being and becoming of logical formalism of epistemological ideas must be articulated through significant scientific and public discourse. We have defined ontology in an engineering sense rather than in the metaphysical sense.

Gruber (1993) explained the meaning of ontology in the engineering sense as, the reality of a concept, relation or fact premised on epistemological roots. The concept of ontology as an analytical theory of determining relationship is used by scientists to explain the process of formation of such functional relations among corresponding entities.

A definition of ontology that comes nearest to our usage is also found in Sztompka (1991, p. 51) who quotes Lloyd (1988, p. 34): "It is the task of science alone to reveal the general, hidden, structural features of phenomena, and the underlying mechanisms of their *becoming*."

Thirdly, the epistemological and ontological levels must be encapsulated in capability and functioning. This level is called the level of evidences, which Heidegger (1988) called Ontic. The concept brings out the analytical, quantitative and empirical policy-theoretic study followed by inferences, policy analysis and recommendations, program formulation and the like. At every point and phase of awareness building in the universal paradigm (UP) there is that indispensable relational causality between the observations on attained states of the entities in given embedded systems and the role of institutions to guide fresh transformation in preferred directions of attaining unity of knowledge in the problem under study.

The E (epistemological)-O (ontological)-O (ontic) phases flow incessantly and continuously, as knowledge formation and its induction of systemic transformation in the light of unity of knowledge are articulated by the intrinsic law of divine unity. But at the end of every such phase through the interconnection of states of the system under investigation and the institutional guidance, there comes about the automatic evolution into new E-O-O learning. The institutional post-evaluation of the degree of unity of knowledge gained in previous experience, which is simulated by a well-defined social wellbeing function of the attenuating problems, charts the new paths of evolutionary learning. At the end of a process and the commencement of a new one there must once again be the recalling of oneness and its subsequent follow-up by the ontological and ontic phases. A full process is thus completed by means of E-O-O-E (evolution coinciding with a recall of epistemology) followed by many such phases of relational learning in unity of knowledge.

The realization of the universal paradigm (UP) under unity of divine knowledge (laws) in the world scene and the socio-scientific milieu, requires vindication of the methodology so defined. This must be reinforced further by its

209

proven results and public understanding. The last one is a matter of positive policies and can best commence at the educational policy levels.

The question then is this: Will UP suddenly replace the existing science as a revolution and make the latter a normal science? There is no clear answer. The increasing conviction and commitment in the new worldview will be a factor of the speed of transformation. The proof of the system in the realms of thought, behavior, implications and applications will matter. Much of the existing attitude of science towards the UP of the divine laws must change in a positive direction of accommodation and understanding. Regretfully, modern science has assumed a hostile climate of opposition to God in favor of materialism (Dampier, 1961). This must change by accommodative will. Hence there will be a great role for positive discourse and understanding.

In the global scene, the enabling of the positive socio-scientific thinking will depend on wider spectrum of dialogues, rather than being preoccupied with preconceived ideas on science, religion, culture, regions and beliefs. In this book according to the dynamics of the E-O-O-E process worldview we promote a climate of global dialogue between civilizations in opposition to the mistaken idea of global clash of civilizations (Huntington, 1993).

Yet in the end, it will be a fact that global transformation will be incremental in nature. In the worst case, this experience can ultimately end up in bifurcated understanding of the world-systems, with one based on unity of knowledge and the other on linear differentiated and disenfranchised, individuated perspectives of socio-scientific opposites. If learning of whatever kind is kept alive in all civilizations then there will exist at least the impact of ideas from the unity worldview on transformations of the differentiated world-systems (Holton, 1992).

In the global scene, awakening from a long bondage of ideologies, it might even be that mankind will commence making subtle movements incrementally. "The process can be likened to subtle movements required to dismantle one of those puzzles — you have to find a spot to make a tiny move and then find another spot to make another tiny move and so on until, eventually, the cube breaks open into its component parts and is ready to be put together again. In the same way, we have to unlock, bit by bit, the rigid structures that have formed in our minds and thus resort to the enlightened worldview of unity of knowledge premised on the intrinsic law that formalizes all of socio-scientific thought and experience". (Shakespeare & Choudhury, 2006).

Structuring the E-O-O-E Worldview of Unity of Knowledge

As we mentioned above the E-O-O-E is an automatic structure that is neither imposed nor concocted. It is natural and invincible to intellection. That is because any thought must rely firstly on a premise. If the premise to be uniquely chosen is of unity of knowledge, for the moral, social and ethical embedding, then knowledge, life and experience must be premised on a relevant premise of knowledge. This is the meaning of Epistemology, the theory of knowledge that identifies and configures how a body of knowledge is derived and organized to attend to any set of issues, problems and questions within embedded systems (Smith, 1992).

Character of the Universal Epistemology

The most important problem of discerning the selection of Epistemology is to find the law and text that most universally organize the worldview of unity of knowledge. The question stands: Can received philosophy of science establish the worldview of unity of knowledge? We will now answer this question from the socio-scientific and moral points of view.

The study of the entire body of knowledge of the Eastern and Occidental world-systems shows that the reduction of human thought relegates to man alone. That is, Reason is the ultimate arbiter of knowledge, and God, though acknowledged for worship in many world-systems, yet remains outside human experience. Even when a claim of socio-scientific and moral association with God is maintained, there is no means of transmission from God to the world-systems through the medium of the divine laws. Consequently, the reality of God remains subjectively dependent on human feelings. God remains a private enterprise. The primal role of divine unity of knowledge and its capability and functioning enabled by the catalysis of a medium other than the subjectivity of human rationalism, remained impossible.

This is the case with the recent intellection on complexity, rationalism and post-modernist epistemology of Giddens (1983), Wallerstein (1998), Heidegger (1988), Husserl (1965) and the entire school of neoclassicism and its genre in mainstream economics (Phelps, 1989; Zsolnai, 2002) and of science as process (Darwin, 1936; Prigogine, 1980; Popper, 1972; Hull, 1988; Dawkins, 1976). The message in all of these ways of understanding the origins of knowledge is subjectivity of human rationalism. The result in the philosophy of science including the moral law, the social order including economics and politics, and the natural sciences, was a conflicting and differentiated understanding of human experience. The impact was felt equally on the hegemony of science over culture and traditions and the political and technological power against the pursuit of truth.

The consequence of such power-shifts was the birth of colonialism that governed and taxed the resources of India to fuel the industrial revolution of

Europe (Heilbroner, 1985). Marx's over-determination epistemological theory vouched for a theory of permanently disequilibrium and conflicting world-systems (Resnick & Wolf, 1987). Technology became the instrument of transference of the development, educational and political models from the West to the rest of the world (Todaro and Smith, 2006).

During the Eighteenth Century European Enlightenment the origin of knowledge was equally premised on human rationalism. The works and beliefs of the scholastic thinkers like Aquinas (1946), Kant (1964) and Hume (1988) reflect the incapability of God to be projected in a functional and capacitating way through logical formalism into the living world-systems and the subtle socio-scientific experiences.

Carnap (1966) wrote on Kant's problem of heteronomy as the dichotomy entailing the following kinds of realities: Pure Reason versus Practical Reason; *a priori* versus *a posteriori*; noumena versus phenomena; the sensate versus the intelligible; moral imperative versus the sensate world. These are examples of the nature of dichotomy that marred the possibility for the divine laws and its spanning relations to be mapped onto the world-systems. This is despite that God was revered by Kant as the highest existent and the source of the moral imperative. The problem of rationalism is the gap in knowledge caused by the absence of a transmission medium for mapping Pure Reason as the moral imperative onto Practical Reason; that is the noumena onto phenomena, so as to comprise a unified and a comprehensive phenomenology of knowledge. The problem of rationalism thus arises from that of Kant's heteronomy.

Incidentally, even the Muslim scholastics earlier were under the same influence of rationalism. The famous Islamic scholar and epistemologist, Abu Hamid [Imam] Al-Ghazali wrote (Marmura, 1997, p. 107, edited) on the problem of the rationalist Islamic philosophers: "What is intended is to show your

213

(rationalists) impotence in your claim of knowing the true nature of things through conclusive demonstrations, and to shed doubt on your claims. Once your impotence becomes manifest, then [one must point out that] there are among people those who hold that the realities of divine matters are not attained through rational reflection – indeed, that is not within human power to know them. For this reason, the giver of the law has said: 'think on God's creation and do not think on God's essence.'"

Structure of *Tawhidi* String Relation (Oneness of God = Unity of Divine laws)

Expression 7.1 delineates the *Tawhidi String Relation* (TSR) of Unity of Divine Knowledge (Tawhid) as the total phenomenology comprising E-O-O-E phases existing in continuous processes of learning. Expression 7.2 further extends the TSR to multi-dimensional space with knowledge induction of the systemic entities.

The following symbols are defined:

Ω denotes the super-cardinal topology of the fullness, perfection and absoluteness of knowledge premised in the divine laws. In the Islamic epistemological meaning this super-cardinal topology comprises the source of the Qur'an in its primal form of completeness.

The concept of topology is that it is a mathematical entity that encompasses all forms of combinations of relations connecting things of the same or opposite types, including the limiting case of including both itself and the nullity. Ω serves as the open cover of all such included subsets, subspaces and relations (Maddox, 1970).

Ω is referred to as the super-cardinal topology because of its openness and super-encompassing nature with the power to explain all realities through relations of such entities with the divine laws. It is therefore neither possible nor necessary to measure super-cardinality of Ω. The true, necessary and sufficient condition for Ω is to generate open sets of extensive relations based on the divine laws. Such relations by generating knowledge-flows that induce the cognitive and material entities and relations carry the principle of pervasive complementarities most extensively. Ω is never comprehended fully by human apprehension, because of its super-cardinality and extensions of mappings of relations. These properties of Ω convey the dynamic nature of the generated knowledge-induced relations between complementary entities.

The veil of ignorance in human vision to know the fullness of Ω requires mapping of quanta of knowledge-flows extracted from the divine origin as the epistemology. This transmission of knowledge-flows contained in Ω onto the human order is realized by the existence of an intermediary denoted by 'S'. What S is in relation to Ω, marks the key point of understanding the difference between the resulting discursive model of unity of knowledge contrary to that of rationalism. We have explained above that if S is premised on rationalism, that which Heidegger and Husserl referred to as Dasein, then human subjectivity will enter the mechanism of deriving knowledge from an empty background that is never complete. As we explained earlier, this absence of completeness and the presence of human subjectivity in the project of rationalism cannot fathom either the divine laws as substantive or connect the divine laws through well-defined relations with the world-systems.

In Islamic epistemology the transmission mapping 'S' must be of such a nature that it remains unsullied by human vagaries of rationalism, and must be capable of extracting the divine roots in a unique way for use by relations and explanations of world-system issues. 'S' therefore denotes the guidance of the Prophet Muhammad, whose sayings and practices comprising the Sunnah are the

215

only known record of prophetic guidance in the annals of religions to date. 'S' extracts the divine laws from the primal source of Ω in the form of rules for further discourse, explanation and application in the world-systems.

The tuple (Ω, S) forms the fundamental Islamic epistemology of divine unity, Tawhid, and whose texts are the Qur'an and the Sunnah (guidance of the Prophet Muhammad). In the realm of mathematical logic, the confirmation of (Ω, S) by itself, as in the case of the irreducible nature of divine truth of oneness of God reflected in the epistemology of unity of knowledge, represents the basis of self-referencing (Godel, 1965). In other words, (Ω, S) will recur in every problem delineation and problem-solving across nexus of domains and continuums. This is a phenomenon of the transmission mechanism that we refer to as 'recalling' of (Ω, S).

$\{\theta\}$ denotes the set of knowledge-flows derived from the epistemology of (Ω, S) through discursive mappings involving conceptualization, methodology and formalism of the system upon which explanation and solving of problems depend. This is the phase that firstly projects the state of the system. The states are then studied and alternatives selected by institutional measures for reforming the state variables as needed along the direction of unity of knowledge as premised on and articulated by (Ω, S).

This kind of conceptualization by examining the existing state of things is the phase marking the ontological principle. It leads to formalization and theory building. In concert with our earlier definition, the meaning of ontology here is the process of *being to becoming* of a body of intellection and concepts that emanate from the rules extracted from the basis of (Ω, S) through the socio-scientific discursive process.

$\{X(\theta)\}$.comprises the vector, matrix and tensors formed simultaneously as entities by the impact of embedded knowledge so to give these entities their meanings. Leibniz (see Russell, 1990) in European Enlightenment and Imam Shatibi (Masud, 1994) on the side of Islamic scholasticism understood the concept of 'meaning' of things in terms of certain forms of their relations in the domain of entities. Shatibi understood the concept of 'meaning' in the sense of extension of knowledge shared between entities according to their responses to the divine laws. This in essence is conveying the epistemology of unity of knowledge in the divine laws. Leibniz, a firm believer in the existence of God did not believe in extension of relations between entities. He considered each entity to be independently endowed by its soul-like behaviour, which Leibniz argued was primordially endowed by God.

In the epistemology of unity of divine knowledge carried by (Ω,S) on to the world-systems created by the spanning of knowledge-flows and their induced entities, the tuples denoted by $\{\theta,X(\theta)\}$ form the domain of formal and cognitive existence as designed by the divine laws of oneness of God in the scheme of *all* things. Such a natural design of oneness in reality is referred to in the Qur'an as the worshipping of God by everything. Some of these members for our understanding of the socio-scientific order are specific entities. This intrinsic worshipping is referred to as the dynamic experience of worshipping, 'Tasbih', in everything. The essence of recognizing the natural law of unity of the divine laws in everything is referred to as divine essence, 'Fitra'. Thus the world-systems, referred to in the Qur'an as Alameen, are continuously immersed in the worshipping of God's oneness through their essence and in response to the primordial oneness of the divine laws that forms the world-systems.

Along with the conceptualization and delineation of the domain, $\{\theta,X(\theta)\}$,

217

pertaining to specific issues and problems under investigation, also come about the formalism underlying the study of such thematic issues. This is the logical formalism that rigorously externalizes the law of divine unity into specific themes by means of relational epistemology between the learning entities of $\{\theta, X(\theta)\}$. The relations extend from simple ones to complex ones. A mathematical or a discursive formal model thus develops at this first conceptual level carrying the 'meaning' of oneness of the divine laws in terms of the primordial epistemology of (Ω, S) through the discourse medium of deriving knowledge-flows $\{\theta\}$ from the premise of (Ω, S).

Formalism of Unity of Divine Knowledge

The sequential relationship denoted by the topological mappings between the respective inter-state relations denoted by $\rightarrow$, is denoted by the chain, $(\Omega, S) \rightarrow \{\theta\} \rightarrow \{\theta, X(\theta)\}$.

At the end of this one-directional causation the evaluation of the existentialist thematic problem is analyzed. This is done by converting the consequences of the state-institutional interrelations based on an existing experience in unity of knowledge vis-à-vis the problems under study. The criterion for such measurement is the method of simulation, that is learning in the domain of $\{\theta, X(\theta)\}$, where $\{\theta\} \in (\Omega, S)$.

There is a further refined relation here that makes only Ω as the primordial and S as transmission interconnecting Ω with $\{\theta, X(\theta)\}$. Thereby, $(\Omega \rightarrow S)$ taken exogenously in the sense of 'recalling' continuously and across continuums of systems, cause the ontology of the world-system to be spanned by $\{\theta, X(\theta)\}$.

Now at the end of the one-directional causation denoted by, $[(\Omega \rightarrow S) \rightarrow \{\theta\} \rightarrow \{X(\theta)\}]$, a criterion function is evaluated by simulating the experience in unity of knowledge realized by such existing relationships. This is done by simulating the criterion function called the Wellbeing Function $(W(\theta,\mathbf{X}(\theta)))$ in the $\{\theta,\mathbf{X}(\theta)\}$-variables, subject to the circular causation between these state-institutional and behavioural variables. Since $\{\theta,\mathbf{X}(\theta)\}$ denotes such state-institutional-behavioural variables that are unified together, the way to explain such unifying interrelations is through circular causation between the variables. This part of the formal simulation model assumes an elaborate structural form amenable to estimation by statistical and coefficient-perturbation methods (Choudhury & Hossain, 2006).

The above chain relation is now extended to the form,

$$(\Omega \to S) \to [\{\theta\} \to \{\mathbf{X}(\theta)\} \to \text{Simulate}_{\{\theta\}} \ W(\theta, \mathbf{X}(\theta))] \qquad (7.1)$$

Subject to, structural set of circular

Causation between the $\{\theta, \mathbf{X}(\theta)\}$-variables

Estimating and changing their unifying

Interrelations by computer-assisted

Methods.

The bracketed sequence [.] in expression (7.1), forms a process of learning. It comprises conceptualization, formalism and one-process evaluation of the entire phenomenological model of Tawhidi (of unity of the divine laws) consciousness in respect to real problems and issues under investigation. The $\{\theta, \mathbf{X}(\theta)\}$-variables within [.] learn continuously and across continuums in respect to the epistemology of $\{\theta\} \in (\Omega \to S)$. $(\Omega \to S)$ is primordial, and hence exogenous, but is continuously 'recalled' in the on-going processes, as shown below. On the other hand, the unification of knowledge by continuous learning across continuums of interrelating systems establishes the multivariate and

219

multidimensional knowledge-induced relations to become of the endogenous type.

The completion of the simulation phase of expression (7.1) that closes a one-process evaluation of the degree of consciousness in the systems, problems and issues under investigation, is referred to as the ontic (Heidegger, 1988; Sherover, 1972). This part of the integrated model marks the evidential phase in process 1. What is true of Process 1 is repeated across simulated levels of $\{\theta, X(\theta)\}$-variables in subsequent evolutionary processes caused by learning in unity of knowledge.

The recalling of the epistemology of Tawhid (unity of the divine laws) causing evolutionary regeneration of processes like (7.1) is shown in expression (7.2).

$$[\text{Process 1}] \qquad\qquad \text{Evolutions: } [\text{Process 2}] \to \dots \tag{7.2}$$

$|$ $\downarrow$

$(\Omega \to S) \to [\{\theta\} \to \{X(\theta)\}$ $\text{Simulate}_{\{\theta\}}\ W(\theta, X(\theta))]$ $\text{Recalling} \to$

          $\underline{\quad\downarrow\quad}$ Subject to, structural set of   Continuity &

     $\{(\theta, X(\theta)\} \to \{$Circular causation between   across continuums

                the $\{\theta, X(\theta)\}$-variables

The evolutionary learning processes of expression (7.2) along simulated new levels of consciousness cause refined concepts and applications of the $\{\theta, X(\theta)\}$-variables to world-system issues and problems of specific kinds. Yet this kind of evolutionary learning dynamics is permanently premised on the exogenous law of unity that determines the entire character of the creative dynamics.

Yet to endow the evolutionary processes with this character of unification it is essential that the chain of processes in expression (7.2) must end up in the terminal cumulative universe of perfect unification of knowledge along with the optimal state variables. Since this state of the evolutionary system cannot be attained by optimization under any condition in the small and large scale universes in temporal order, therefore the only possibility for the closure is to attain the universe of Ω itself.

The necessary and sufficient condition of attaining unity of knowledge is the realization of the summarized universal relation shown in expression (7.3). This expression represents the complete phenomenological model of the Tawhidi genre.

The Tawhidi Phenomenological Worldview

$$(\Omega \to S) \to \text{Evolution: Cumulatively Learning World-System } \{\theta, X(\theta)\} \to \Omega$$

$$(7.3)$$

The necessary and sufficient essentiality of expression (7.3) in establishing the complete phenomenological worldview as the Universal Paradigm can be readily deduced. The singular axiom of every problem discussed within expression (7.3) is that the world-systems are centred in unity of knowledge whose epistemology emanates from the foundation of $(\Omega \to S)$. All other aspects of TSR are derived consequences of the Tawhidi axiom.

1. The Necessary Condition

Firstly, we argue by assuming that expression (7.3) is true. This is the necessary condition. The implication then is that the simulation of the evolutionary learning system cannot end in finite continuums. Consequently, there is no optimum in any real-world problem-solving situation. There are only acquired states of the problem-solving systems, as caused by simulation in given processes according to the methodology that was substantively explained above by means of the E-O-O-E worldview. Optimum in a continuously learning universe across its continuums of interrelating and unifying systems is now possible either in the instantaneous core of the system or in the closures of Ω. Such end-of process closure marks the completion of all knowledge-flows and the attainment of the optimal blessing of $\{\theta, \mathbf{X}(\theta)\}$ in the form of wellbeing. The Qur'an refers to this Event of the End as Akhira (the Hereafter). The optimal blessing caused by such closure of $\{\theta, \mathbf{X}(\theta)\}$ is referred to as the Supreme Felicity. We note that the same kind of optimality was extant in primal creation. That is in the divine laws. But the primordial Ω remains unobserved. The Qur'an calls this primordial state as the Event of the Beginning.

Optimality of the instantaneous case is impossible and meaningless, for in the state of continuous learning across matter-mind domains of reality we cannot hold any interacting relation to remain constant. Quantum Physics has proven this fact for the sub-atomic worlds.

The other case is of the cumulative state of the End as of the primordial state of the Beginning. By definition of the Hereafter in simulation calculus the dimension of Ω is now treated as super-cardinal. Super-cardinality denotes the non-configurable dimensionality of the primal Ω, which is actualized only in the Hereafter. Consequently, only relations between attained $\{\theta, \mathbf{X}(\theta)\}$ are now possible in the end states of Ω. No further evolution can occur in the super-cardinal states of Ω. Therefore, the wellbeing function $W(.)$ is optimized by the completion of unity of knowledge across the entirety of unravelling processes of divine unity overarching the super-cardinal domain of Ω. The equivalence of Ω is in respect of either the End or the Beginning.

2. The Sufficiency Condition

For proving the sufficiency condition, we argue as follows: Let there be a tuple $\{\theta, \mathbf{X}(\theta)\}$ belonging to the TSR (expression (7.3)). Then such a tuple is governed by the law of unity of knowledge. All the simulation conditions apply. If possible now, let a process-based simulation state end prior to Ω of the Hereafter (= Primal). Then an optimum for $W(.)$ is attainable under the simulation relations. Consequently, Ω remains unrealized at the End. Likewise, Ω remains undefined in the Beginning, since expression (7.3) is then open with respect to Ω at the End. Thus, a truncation is now required to define the tupe $\{\theta, \mathbf{X}(\theta)\}$ in respect to optimizing $W(.)$ in a truncated domain within expression (7.3) not including Ω in the Beginning and the End. $\{\theta, \mathbf{X}(\theta)\}$ cannot therefore be well-defined within expression (7.3) as given. We thereby arrive at a contradiction. Therefore, in order for $\{\theta, \mathbf{X}(\theta)\}$ to exist with respect to expression (7.3), the TSR must be attained.

Critical Properties of the Circular Causation Relations

The process-oriented continuity of expression (7.2) completes the E-O-O-E Tawhidi model of unity of divine laws (unity of knowledge). There are a number of critical properties of learning in the interrelations of the circular causation structural relations of expression (7.2). We now explain these properties of evolutionary learning in unity of knowledge. We refer to expression (7.2).

$\{\theta, \mathbf{X}(\theta)\}$-variables in the various mathematical definitions represent the tuple combining state variables and institutional and behavioral variables. This combination necessitates that the tuple is formed by extensive discourse (Interaction) between the realities of the socio-scientific systems and the institutional-behavioral agencies. The method towards a final determination of consensus (Integration) to realize a limiting value of knowledge-flow from the discourse sequences of knowledge-flows in any given learning and evaluation process goes through the institutional discourse to derive such limiting knowledge values indicating levels of attained consciousness in understanding the rules derived from the epistemology of (Ω, S) and simulating it thereby to higher levels of consciousness.

The ability to arrive at such rules involves evaluation that is done by the simulation method using circular causation between the interacting variables in the hope of transforming the system towards higher levels of consciousness in unity of knowledge. This part involves the ontic study of formal system. Up to the point of logical formalism of evaluation we attain Interaction leading to Integration of the learning process in response to knowledge-flows premised on (Ω, S). Subsequent to every process of learning involving Interaction leading to Integration in the small and large scale universes represented by specific problems, there is fresh learning that re-appears in evolutionary phases, which repeat the processual experience. Hence the properties of learning in Tawhidi epistemology comprise *Interaction* leading to *Integration* leading to *Evolution* (IIE) in continuums. These properties of learning apply to every particular socio-scientific world-system in view of the definition of an UP proposition and application to specific issues and problems under investigation. Consequently, systemic unification on the basis of unity of divine knowledge is universally characterized by IIE-processes. The Tawhidi worldview in its ontologically applied form now results in the IIE-process-based methodology.

The final delineation of the IIE-processes in view of the E-O-O-E parts of the TSR is this:

Epistemology (E) →   Ontology (O) → Ontic (O) → Evolutionary (E)
Non-Process Exogeneity   Process-Based Endogeneity by Learning
(Ω,S)   →   {θ,X(θ)} →   Evaluation → Continuity
                    Interactive, Integrative and Evolutionary Field (IIE)   (7.4)

Proof of the Universal Paradigm Project in TSR

The project of phenomenology as the study of consciousness reaches its ultimate height in an overarching knowledge enterprise when it has its unique capability to deliver certain principal prospects. These comprise the capability to generate well-defined laws, rules, relations and inferences premised on the most irreducible foundation of oneness of God (unity of knowledge = Ω). The proof of such irreducibility follows the principle of self-referencing (Choudhury, 2002). The project must be capable of the widest possible coverage to apply its methodology of the worldview.

Also the same unique worldview methodology must address both the positive knowledge framework as well as the negative 'de-knowledge' framework. This signifies, as we have explained earlier that, the 'meaning' of Truth and Falsehood both ensue from the same unique and irreducible epistemology of Tawhid. The above points are clearly proven by the TSR.

Self-referencing is proven by the fact that continuing processes in expression (7.2) repeat throughout the entire evolutionary learning process in reflexive interrelationships between deductive and inductive reasoning. The epistemology of unity of knowledge premises the seat of laws, rules and

225

behavioural traits on specific texts and their institutional interpretation by discourse. These foundational texts are the Qur'an and the Sunnah.

Interpretation of the rules derived from the texts is done by discourse (IIE-process) by exercising the method of interaction and integration or consensus. The same method of primal reference to the Tawhidi epistemology, its irreducibility and inferences derived there from by interpretive and discursive experience, forms the core and the evolutionary periphery, respectively.

Extensive application of the Tawhidi worldview methodology is gained by realization of the entire E-O-O-E Process. The polity-theoretic combination of such an experience measures the learning dynamics in unity of knowledge by using the circular causation ontic method. Such a method, which arises from the TSR is applicable to an increasing range of socio-scientific concepts, models issues and problems, as it continues on to provide the necessary and sufficient conditions of deductive and inductive reasoning. The necessary and sufficient conditions of TSR in reference to Ω were proved above.

Explaining 'De-Knowledge' by TSR

The explanation of 'de-knowledge' (Choudhury, 2000) is done in the same way as for 'knowledge'. Though in this case, rationalism, methodological individualism and Darwinian kind of mutation between locally limited interacting entities, lead to bundles of bifurcated forms. Thus the totality that drives the de-knowledge domain in respect to specific issues, problems and concepts comprise continuously bifurcating relations between entities (Choudhury, 1999). In reference to the Tawhidi worldview, such kinds of properties of differentiated world-systems and their specific concepts, problems and issues are well-defined by the divine laws in terms of denying the Qur'anic principle of pairs (principle of pervasive complementarities).

The three categories of understanding regarding Truth, Falsehood and Indeterminate are well-defined in the Qur'an and are explicated by the Sunnah. Some of the verses of the Qur'an clearly state what is recommended and what is forbidden, and thus the differences between Truth and Falsehood. Other verses are silent in some instances. In these cases well-defined determinateness between Truth and Falsehood emerge by the process of learning and the advance of consciousness.

CONCLUSION

The worldview of the Universal Paradigm and its methodology were developed here in the framework of unity of divine knowledge. In the Islamic case, the universal paradigm is explained by the epistemology of Tawhid. This chapter has brought forth the epistemological methodology and details in the Islamic context. The chapter has also proved how the Tawhidi worldview is a supra- revolutionary doctrine and paradigm. In this direction towards establishing the Tawhidi worldview as the universal paradigm, comparative and contrasting examinations of other epistemological ideas in socio-scientific thought were studied. The critical examination of these ideas led to the question on why and how the Tawhidi worldview stands out to be the universal paradigm in the sense of the worldview being substantively explained in contrast to normal science, paradigm and scientific revolution.

REFERENCES

Aquinas, T. 1946. "The existence of God in things", The infinity of God", "The eternity of God", "The unity of God", "Of God's knowledge", in *Summa Theologiae*. Vol.1, pp. 1-11, 30-34, 40-45, 46-48, 72-86, New York, NY: Benziger Brothers, Inc.

Carnap, R. 1966. "Kant's synthetic *a priori*", in his *Philosophical Foundations of Physics*, Ed. M. Gardner, New York: Basic Books, Inc.

Choudhury, M.A. 1999. "Globally interactive systems", in his *Comparative Economic Theory, Occidental and Islamic Perspectives*, Chapter 3, Norwell, MA: Kluwer Academic.

Choudhury, M.A. 2002. *Technical Appendix: The Methodology of Self-Referencing in Scientific Reasoning, in Explaining the Qur'an, A socio-Scientific Inquiry*, Chapter 4, Lewiston, New York: The Edwin Mellen Press.

Choudhury, M.A. 2000. *The Islamic Worldview, Socio-Scientific Perspectives*, London, Eng: Kegan Paul International.

Choudhury, M.A. & Hossain, M. S. 2006. *Development Planning in the Sultanate of Oman*, Lewiston, New York: The Edwin Mellen Press.

Dampier, W.S. 1961. *A History of Science and Its Relations with Philosophy and Religion*, Cambridge, Eng: Cambridge University Press.

Darwin, C. 1936. *Descent of Man (also known as The Origin of Species)*, New York, NY: Modern Library.

Dawkins, R. 1976. *The Selfish Gene*, New York: Oxford University Press.

Giddens, A. 1983. *A Contemporary Critique of Historical Materialism, Vol. 1: Power, Property and the State*, particularly pp. 26-48, Berkeley, CA: University of California Press.

Godel, K. 1965. "On formally undecidable propositions of Principia Mathematica and related systems", in M. Davies Ed. *The Undecidable*, New York: Raven Books.

Gruber, T.R. 1993. "A translation approach to portable ontologies", *Knowledge Acquisition*, Vol. 5, No. 2, 199-200.

Heidegger, M. trans. A. Hofstadter,1988. *The Basic Problems of Phenomenology*, Bloomington & Indianapolis, IN: Indiana University Press.

Heilbroner, R.L. 1985. *The Nature and Logic of Capitalism*, New York: W.W. Norton

Holton, R. J. 1992. "Economic liberalism and the theory of the market", in *Economy and Society*, London, Eng: Routledge, pp. 52-69.

Hull, D.L.1988. "Science as a selection process", in his *Science as a Process*, Chapter. 12, Chicago, IL: The University of Chicago Press.

Huntington, S.P. 1993. "Clash of Civilizations?" *Foreign Affairs*, Summer.

Hume, D. 1988. *An Enquiry Concerning Human Understanding*, Buffalo: Prometheus Books.

Husserl, E. trans. Q. Lauer. 1965. *Phenomenology and the Crisis of Philosophy*, p. 155, New York: Harper & Row Publishers, particularly note his comment on the perceived absence of ethical roots in occidental scientific inquiry.

Kant, I. trans. H.J. Paton. 1964. *Groundwork of the Metaphysics of Morals*, New York: Harper & Row Publishers.

Kuhn, T. 1970 [reprinted]. "The Structure of Scientific Revolutions", in O. Neurath, R. Carnap & C. Morris eds. *Foundations of the Unity of Science* Volume II, Chicago, ILL: University of Chicago Press, pp. 53-271.

Lloyd, C. 1988. *Explanation in Social History*, Oxford, UK: Basil and Blackwell.

Maddox, I.J. 1970. *Elements of Functional Analysis*, Cambridge, Eng: Cambridge University Press.

Marmura, M.E. (trans.) 1997. *Al-Ghazali, The Incoherence of the Philosophers*, Provo, Utah: Brigham Young University Press.

Masud, M.K. 1994. *Shatibi's Theory of Meaning*, Islamabad, Pakistan: Islamic Research Institute, International Islamic University.

Phelps, E.S. 1989. "Distributive justice", in J. Eatwell, P. Newman & M. Milgate eds. *New Palgrave: Social Economics*, New York, NY: W.W. Norton.

Popper, K. 1972. *Conjectures and Refutations: The Growth of Scientific Knowledge*, London, Eng: Routledge & Kegan Paul.

Prigogine, I. 1980. *From Being to Becoming*, San Francisco, CA: W.H. Freeman.

Resnick, S.A. & R.D.Wolff. 1987. *Knowledge and Class, A Marxian Critique of Political Economy*, Chicago, IL: The University of Chicago Press.

229

Russell, B. 1990. "Leibniz", in *A History of Western Philosophy*, London, Eng: George Allen & Unwin,.

Shakespeare, R. & Choudhury, M.A. (ms) 2006. *The Universal Paradigm*, mimeo. Postgraduate Program in Islamic Economics and Finance, Trisakti University, Jakarta, Indonesia.

Sherover, C.M.1972. *Heidegger, Kant and Time*, Bloomington, IN: Indiana University Press.

Smith, T.S. 1992. *Strong Interactions*, Chicago, IL: University of Chicago Press.

Sztompka, P. 1991. *Society in Action, The Theory of Social Becoming*, Chicago, IL: The University of Chicago Press.

M.P. Todaro & S.C. Smith, 2006. *Economic Development*, Toronto, Ont: Pearson

Wallerstein, I. 1998. "Spacetime as the basis of knowledge", in O.F. Borda Ed. *People's Participation, Challenges Ahead*, pp. 43-62, New York: The APEX Press.

Zsolnai, L. 2002. "The Moral Economic Man", in his *Ethics in the Economy*, New York, NY: Peter Lang, pp. 39-58.

TECHNICAL APPENDIX TO CHAPTER 7: SIMULATION BY MEANS OF CIRCULAR CAUSATION RELATIONS

Refer back to expression (7.2) in Chapter 7:

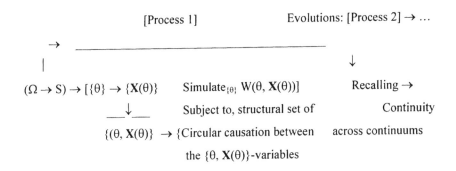

[Process 1]   Evolutions: [Process 2] → ...

$(\Omega \to S) \to [\{\theta\} \to \{X(\theta)\}$ Simulate$_{\{\theta\}}$ W(θ, **X**(θ))]   Recalling →

Subject to, structural set of   Continuity

$\{(\theta, X(\theta)\} \to \{$Circular causation between   across continuums

the $\{\theta, X(\theta)\}$-variables

$(\Omega \to S)$ is exogenously given and is recalled continuously to generate the knowledge-flows $\{\theta\}$ and simultaneously the knowledge-induced $\{X(\theta)\}$ across processes. Thus comes about $\{(\theta, X(\theta)\}$ after a series of finite number of institutional interaction to arrive at a consensual value denoted by θ. Consequently, there are specified values attained by the vector $X(\theta)$. Thus the formalism of Interaction and Integration proceeds at this phase. We represent this by the following simulation problem. It is primitively conceptualized at this stage according to the implications of the discourse process premised on the dynamics of discovering a degree of consciousness of unity of knowledge.

$$\text{Simulate }_{\{\theta\}} W(\theta) = W(\theta, X_1(\theta), X_2(\theta), ..., X_n(\theta)) \qquad \text{(A7.1)}$$

$$\text{Subject to, } X_i = f_i (\theta, X_1(\theta), X_2(\theta), ..., X_{i-1}(\theta), X_{i+1}(\theta), ..., X_n(\theta)) \qquad \text{(A7.2)}$$

$$\theta_* = g(\theta, X_1(\theta), X_2(\theta), ..., X_n(\theta)) \qquad \text{(A7.3)}$$

i = 1,2, ..., n.

This stage simply formalizes, without experimentation using the model, a discursive notion of attaining unity of knowledge between the variables (θ, $X(\theta)$). This approach is signified by each of the equations (A7.1) – (A7.3) and all together. Some of the n-number of variables can be policy variables and rules based on understanding (consciousness) denoted by a limiting θ-variable.

231

The limiting value of θ-variable is discoursed by interaction and integration institutionally on the basis of consciousness and understanding. In this way, θ-variable values comprise ordinal assignments corresponding to the observed values of $X(θ)$-values. Consequently, equations (A7.1) – (A7.3) can be estimated as a structural system of regression equations. These equations are more appropriately defined by perturbation coefficients with probability distributions caused by the impact of learning in unity of knowledge (Vanmarcke, 1988).

Furthermore, equation (A7.1) and (A7.3) being similar, this implies that $W(θ)$-values are monotonic to θ-values. Therefore, these equivalent values represent the degree to which unity of knowledge is attained through interaction and integration in the learning system in respect of the problem under study. The definition of the Wellbeing Function, $W(θ)$, is essentially this as a measure of evaluating the experience of the learning system in unity of knowledge through a combination of interrelationships between the state variables of the system under study and the institutional policy and guidance variables derived from the epistemology of $(Ω{\rightarrow}S)$ through discourse represented in determining limiting ordinal θ-value for the problem under investigation (Choudhury & Zaman, 2006).

The following estimation problem is now in place when we come to the ontic (evidential) stage, which follows the above conceptualization stage of explaining the existentialist formalism of the project of unity of knowledge in specific problems under investigation. Since it is impossible to define the 'being' of God in Islamic belief, all that can be done in functional use of the precept of divine oneness is to axiomatically assume the unity of the divine laws and then prove its relational and functional power of explanation. The stage of formalism explained above is therefore the 'being' of the divine laws (not of God, which remains impossible of corporeal representation, as the Qur'an says, "no fair estimate of God will you get".). This is the ontological stage, and its formalism defines the Epistemological – Ontological – Ontic relationship (E-O-O-E). The stage of formalism as the 'being' of the divine laws represents the Ontological Principle.

The Ontic stage is the experimental and post-evaluation stage. It involves actual estimation of the ontological model given by (A7.1)-(A7.3) by the computer-assisted methods of mathematics, statistics and argumentation. The entire sequence in any given process as shown in expression (7.2) of Chapter 7 describes the phenomenology of unity of divine knowledge (Tawhid) and becomes a central problem of the project of Computing Reality.

The entire simulation method is now represented in the following way:

1. Well-define the Wellbeing Function in terms of a central consciousness of unity of knowledge in the problem under investigation. For instance, in the problem of poverty alleviation we can define a product function (Π) with respective elasticity coefficients of Wellbeing to the specific variables ($α_i$, i = 1,2, …, n), given a level of understanding of the problem of poverty alleviation within the

framework of unity of divine laws as derived epistemologically from the sources of the Qur'an, the Sunnah and Islamic discourse..

$$W(\theta) = \Pi_{i=1}^{n} X_i^{\alpha_i}; \; X_i = X_i(\theta); \; i = 1, 2,.., n \qquad (A7.4)$$

$W(\theta)$ can be treated as Poverty Alleviation Index. It can be treated in exactly the same way as θ-values are assigned ordinal values by the interrelationship between state variables and institutional variables. Examples are Chenery-Ahluwalia Index (1974; Torras, 1999) and the Human Poverty Index (UNDP, 2005).

θ is the limiting value of knowledge-flows generated by interaction, integration and evolutionary (IIE) learning experience within each process in reference to the observed degree of complementarities attained between the $X(\theta)$-variables by using the law, rules and directions derived from $(\Omega \rightarrow S)$. Ordinal θ-values are thereby assigned in ascending order of complementarities (unifying relationship) between the $X(\theta)$-variables. More on the assignment of such ordinal values to θ appears below.

Examples of $X(\theta)$-variables are entitlement, empowerment, gender and age distribution, distributional instruments, all taken up in participatory state-polity environment.

The tabulated values can be used to estimate the system of equations (A7.4) in relation to the circular causation relations (A7.2) – (A7.3). The predictor values of $X(\theta)$-variables obtained by using the actual values of the explanatory variables in the estimated equations of expression (A7.2)-(A7.3) are now inserted into the estimated (A7.4) expression to come up with the simulated value of $W(\theta, X(\theta))$, the Wellbeing Function for indicating the degree of unity of knowledge gained in the system under investigation.

Since $W(\theta)$ is a monotonic function of θ, we can take its ordinal values to be the same as for θ. Thereby, both θ and $W(\theta)$ values are identical to each other in the simulation problem. In an alternative case, if $W(\theta)$ identified with a variable for which data exists, then its values can be in terms of such statistically recorded variable. An example is when X_1 denotes per capita income of the poor; X_2 denotes number of poor persons. Then $W = X_1 \times X_2$ denotes entitlement of the poor. Likewise, when X_3 denotes a social instrument expressed as charitable distribution per unit of entitlement, then, $X_1 \times X_2 \times X_3$ denotes total charitable distribution based on the level of entitlement.

233

Tabulation of Observations

$$\underline{X_1 \qquad X_2 \dots \quad X_n \ \theta \ W(\theta)} \tag{A7.4}$$

1   x_{11} $x_{12}\dots$ $x_{1n} \ \theta_1 \ w_1$ (x_{1j} complementarities assign ordinal value to

$$(\theta_1, w_1)$$

2   x_{21} $x_{22}\dots$ $x_{2n} \ \theta_2 \ w_2$ (x_{2j} complementarities assign ordinal value to

$$(\theta_2, w_2)$$

3   x_{31} $x_{32}\dots$ $x_{3n} \ \theta_3 \ w_3$ (x_{3j} complementarities assign ordinal value to

$$(\theta_3, w_3)$$

.m x_{m1} $x_{m2}\dots$ $x_{mn} \ \theta_m \ w_m$ (x_{mj} complementarities assign ordinal value to

$$(\theta_m, w_m)$$

$j = 1, 2, \dots, n$

Once the estimation is done and the predictor values denoted by say $X^{\wedge}(\theta^{\wedge})$ are computed from the sets of relations (A7.2-A7.3 & A7.4), we obtain the process-specific simulated value of wellbeing. This is denoted by,

$$W^*(\theta^{\wedge}) = \Pi_{i=1}^{n} X^{\wedge}_i{}^{\alpha i}; \ X^{\wedge}_i = X^{\wedge}_i(\theta^{\wedge}); \ i = 1, 2, .., n \tag{A7.5}.$$

This value of W conveys the degree to which unity of knowledge is attained in the IIE-process 1. Subsequent process estimation proceeds in one of several ways. Firstly, the estimated coefficients of equations (A7.2-A7.3 and A.7.4/A7.5) can be simulated by observing new values that they can assume under policy impacts that change these coefficient values. The policy implications are implied from possible ones.

The alternative coefficient values can be chosen from a field of such values. The methodology to generate such values is Spatial Domain Analysis. Now new values of $X^{\wedge}(\theta^{\wedge})$, and thereby, new values of $W^*(X^{\wedge}(\theta^{\wedge}))$ are obtained. Hence the simulation over processes is carried out as over continuums.

We note here the meaning of the estimated coefficients in all of the mentioned equations. Note that we are working with the multiplicative form of (A7.2-A7.3) of the form shown by A7.4. In general thereby,

$(dX_i/X_i)/(dX_j/X_j)$

$= a_{ij}$ coefficient of X_j = the elasticity coefficient of X_i in relation to X_j,

$i,j = 1,2,...,n$

The driving principle of pervasive complementarities underlying the precept of unity of knowledge by relational linkages between the variables, policy instruments and organization of the development process (socio-scientific system) in respect to the given problem and issue at hand, causes $a_{ij} > 0$ or less negative value over learning processes. Consequently, simulated changes in the coefficients by means of re-assigning values from the field of numbers generated by Spatial Domain Analysis reflect a combination of state-policy effects. Such simulated coefficients chosen from a field of elasticity numbers and pertaining to specific policies denote learning coefficients of the equation system (A7.2-A7.3 and A7.4/A7.5) in respect to the poverty alleviation problem in our present reference. Consequently, the estimated tuple $(\theta, X(\theta))$-values represent the corresponding simulated state-policy variables of the problem under investigation. Simulated values of $W(\theta)$ are thereby obtained by inserting the new predictor values of the tuple with the new sets of coefficient values.

In the case of Islamic poverty alleviation development agenda, the implementation of wealth tax (Zakat) in combination with the participatory development finance instruments, such as Mudarabah (profit and loss sharing), Musharakah (equity) and Murabaha (cost-plus mark-up) is brought about by knowledge induction in household and social preference formation. This is endogenous process characterization that takes place over evolutionary learning in participatory forms of development objectives. Yet the machinery of Government and institutions is used to realize such a program. Simulated θ-values denote the levels of endogenous learning, which finds its impact on the realization of unity of relations (complementarities) between the variables and the effectiveness of the participatory policies and programs socially chosen to bring about such change from a less participatory or trade-off socio-economic state-polity relationships to a more participatory one.

In the case of poverty alleviation problem, output, entitlement, empowerment, development expenditure and charity are examples of complementary variables in unity of knowledge. This though is not necessarily the case with mainstream economics. Considerable trade-offs can be noticed between output and human development. The latter should be a positive function of entitlement, empowerment, output, development expenditure and charity (Choudhury & Korvin & Seyyed, 2002). The participatory development framework points to a reconstruction of $a_{ij} > 0$, for $i,j = 1,2,...,n$. The neoclassical mainstream framework can show trade-offs. Even when some positive values are shown in the second case, this is not due to consistently structural properties of complementarities in such an economic resource allocation. Positive values can come about by short-run policy injection of the exogenous type, not of the endogenously learning type.

REFERENCES

Ahluwalia, M. and H. Chenery, 1974. "The Economic Framework." In H. Chenery, M. Ahluwalia, C.L.G. Bell, J.H. Dully, and R. Jolly, eds. *Redistribution with Growth.* London: Oxford University Press.

Choudhury, M.A. Korvin, G. & Seyyed, F. 2002. "Discovering micro level trade-off in economic development and studying their fractal character", *Indonesian Management & Accounting Research*, 1:1, pp. 49-70.

Choudhury, M.A. & Zaman, S.I. 2006 forthcoming. "Learning sets and topologies", *Keybernetes: International Journal of Systems and Cybernetics*, 35:1-10.

Torras, M. 1999. "Inequality, Resource Depletion and Welfare Accounting: Applications to Indonesia and Costa Rica", *World Development*, 27:7, pp.1191-1202

United Nations Development Program, 2005. "Inequality and human development, *Human Development Report 2005*, Chapter 5, Oxford & New York: Oxford University Press.

Vanmarcke, E. 1988. *Random Fields*, Cambridge, MA: The MIT Press.

CONCLUSION

This book has researched a rare domain of intellectual inquiry, which after the legacy of Herbert Simon and Jacob Wiener recently and Imam Ghazali in the Islamic classical scholasticism has not been studied. We have argued in this work that, as a result of this absence of this gap in critical thought, no fresh epistemological groundwork has been opened up for intellectual inquiry and its analytical and social applications. Much of contemporary pursuits into analytical empiricism and material manifestations of worldviews, such as, capitalism and globalization, liberalism and markets in differentiated economic systems, have failed to reflect on newer dimensions of human potential.

Consequently, the deeper import of morals and ethics has remained divorced from or only exogenously implicated in socio-scientific intellection, institutionalism and social applications. Policy-making, human preferences and life-fulfilling regimes of socioeconomic development have thus been encumbered by self-seeking models of behavior and optimal resource allocation scenarios. The resulting socio-scientific methodology in the absence of an understanding of the greater perspective of human sharing within the laws and preferences of morality and ethics is unable to answer the profound wellbeing issues of the human family. This comprises social co-existence in a unified, embedded and endogenously induced ethical worldview of existence and all that this holds for wellbeing of all.

The research project of this book opens up that much-needed inquiry into the moral and ethical context for understanding reality and then in applying the epistemological foundations of such an understanding to a purposive computational idea. Thereby, we have argued that machines, computers, markets and artifacts are as much ethically induced entities as are human beings and institutions. That is because of the circular causation that pervades in complex non-linear interrelations existing between the domains of abstraction and human agency.

In the abstract domain abide the variables and ontologies for comprehension, formulation and evidential application to diverse socio-scientific issues and problems that arise. The resulting system of complex circular causation interrelations comprises a nexus of relational learning. We referred to such a worldview of unified relational learning across and within systems in terms of entity and agency interrelations as unity of knowledge arising from a fundamental episteme of oneness. We have characterized such a system of circular causation interrelations as a complex and non-linear one, contrasting these with the linear non-learning system of the optimal and steady-state world-systems.

The resulting methodology thus rested on a study of socio-scientific consciousness underlying methodological intellection of learning world-systems. This kind of epistemological conception resulted in the study of consciousness, which we referred to as the project of phenomenology. It is the complete and overarching understanding of unity of knowledge intra- and inter-systemic learning dynamics.

We referred to the systemic understanding of the comprehensive project of phenomenology towards understanding and applying the precept of unity of knowledge in the presence of complexity and non-linearity as neurocybernetic and system theory. We have gone into a critical study of the literature on matter and mind relations for establishing the groundwork of the neurocybernetic and system theory of phenomenology.

At the end, we note that the project of *Computing Reality* opens up fresh inquiry in the much-awaited area of a knowledge-centered worldview that what constitutes the moral and ethical groundwork of understanding reality. Once thus understood against the appropriate epistemology of unity of knowledge, the context of logical formalism and its empiricism and explanation can be laid out.

Machine logic and computer systems, as of the Turing kind, can be developed as supercomputer entities for solving complex problems of learning systems. Reality is indeed complex by its richness, yet rendering to measurement and explanation in cogent ways.

Computing Reality may thus prove to be of a substantive nature in the field of original research in the phenomenology of complex learning systems that is governed by the episteme of unity of knowledge worldview. *Computing Reality* can thus be of interest and usefulness to the serious students, academics and researchers of philosophy of science and its applied real-world manifestations.

SUBJECT INDEX

reflexivity, 177, 181-82, 184-85, 188, 190-191

risk diversification, 99, 113-14

scientific revolution, 121, 209-10, 229, 231

simulation, 11-16, 20, 62, 82, 88, 97, 99, 114, 124, 156, 177-78, 180-81, 190, 220-26, 233-36

social contracts (social contractarianism), 33, 94-98, 120, 151, 155, 160, 179, 194-96, 204

social Darwinism (Darwinian), 3, 24-25, 44, 87, 104, 106, 108, 115, 172, 182, 227

social justice, 91, 146, 149, 150, 152,

social wellbeing function, 11-12, 114, 191, 207

socio-scientific theory, 1, 3, 40, 51, 158, 203

spatial domain analysis, 58, 235-36

spiritual capital, 20, 95-96, 109, 178

state and control variables, 4, 14, 97

steady-state equilibrium, 93, 97-98, 103, 105, 123

super-cardinality, 185-87, 214, 215, 222-23

supercomputer, 187, 241

sustainability, 11, 14-15, 17, 29, 45, 48-49, 116-17, 146-48, 152-53, 158, 177, 180, 195

Tawhidi String Relation, TSR, 214, 221, 223, 225-26

theory of value, 86, 110-11

topology (topological), 123, 172, 184-86, 197-98, 204, 212-13, 216, 234

trans-empirical, 162-63, 166-67, 169, 183-84, 190, 195-96, 199, 202

unity of knowledge, passim

world-system, 1-10, 16, 22-24, 41-42, 44, 47, 50, 84-89, 94, 101, 103, 105, 108-9, 116, 126-28, 132, 136-38, 140-46, 149, 151, 156, 161, 164, 166, 168, 170, 172, 177, 179-80, 184-86, 202, 204-6, 208-11, 213-16, 218-19, 222, 224, 236

GLOSSARY OF ARABIC TERMS

Akhira	the Event of the Hereafter mentioned in the Qur'an. Also interpreted in this paper as the Moment of completion of relational learning in unity of knowledge.
Alameen	world-systems mentioned in the Qur'an
Fana fi-Tawhid	to subdue oneself in Tawhid
Mudarabah	profit sharing in capital and labor Islamic joint venture
Murabaha	cost-push pricing in Islamic cooperative business contract
Musharakah	equity participation by profit and loss sharing in Islamic business enterprise between capital owners
Qur'an	revealed book to the Prophet Muhammad believed by Muslims as the last of the divine revelations completed as divine guidance for mankind
Sunnah	guidance by the sayings, practices and recommendations of the Prophet Muhammad as explication of the Qur'anic law
Tasbih	dynamic personal experience of worshipping God in the midst of consciousness of the divine laws of oneness
Tawhid	Oneness of God; also interpreted in this work as the unity of the divine laws carried to all experiences

ABOUT THE BOOK

This is an original piece of research in the area of unity of knowledge as the epistemology of neurocybernetics and system theory that explains relationally unified socio-scientific world-systems. *Computing Reality* is thereby a rigorous study in the moral and ethical contexts of socio-scientific issues and how these can be formally investigated using the methodology of unity of knowledge. A great extant of analytical philosophical examination between the Occidental world-system and the precept of unity of the divine laws is undertaken to bring out the unique, universal and extensive application of the latter to all issues. In establishing and formalizing such a methodology of unity of knowledge, epistemological, ontological and ontic (evidential) premises are used. The methodology thus turns out to be analytical and sometime mathematical in nature. Much of the mathematical coverage is placed in the technical appendixes, though some chapters also contain sections in mathematical formalism. This is required to bring out the deeply methodological nature of the project of unity of knowledge as neurocybernetics and system theory as a study of unification between soul, mind and matter.

ABOUT THE AUTHORS

Dr. Masudul Alam Choudhury is Professor of Economics in the Department of Economics and Finance, College of Commerce and Economics, Sultan Qaboos University, Muscat, Sultanate of Oman. He is also the International Chair of Postgraduate Program in Islamic Economics and Finance, Trisakti University, Jakarta, Indonesia. Professor Choudhury retired from the School of Business, Cape Breton University, Sydney, Nova Scotia, Canada. Professor Choudhury is a contemporary pioneer in the areas of epistemological foundations of Islamic socio-scientific thought, applied to Islamic political economy, world-system studies and various areas of Islamic economics and finance in comparative perspectives. His most recent publication comprises five volumes entitled *Science and Epistemology in the Qur'an* (with different volume titles) published by the Edwin Mellen Press, Lewiston, New York, 2006. Professor Choudhury earned his M.A. and Ph.D. Degrees from the University of Toronto.

Dr. Mohammad Shahadat Hossain is Associate Professor and Chairman of Computer Science in Chittagong University, Bangladesh. He specializes in Spatial Domain Analysis of Geographical Information Systems. He has published internationally in this area. His most recent work with Professor Choudhury is entitled *Development Planning in the Sultanate of Oman* (Lewiston, New York: The Edwin Mellen Press, 2006). Dr. Hossain earned his Ph.D. Degree from University of Manchester Institute of Science and Technology.

Other Titles by
blue ocean press

Parables of Milk and Might: Development Political Satire – >>The Voices of the Affected<<

By RAN (2007)
ISBN: 978-4-902837-

Following over four decades of development politics, after the official end of colonialism in most countries in Africa, South America and Asia, it is difficult for the industrial countries to forego their economic interests in the developing countries, which are said to be independent. Their continued presence in these countries, controlling or dictating the trend of economic and political developments, is a proof of the protection of their interests.

Parables of Milk and Might is a satire on the international development sector, in particular, the relationship between the countries of the Global North and South. The book uses a wonderful combination of wordplay, metaphor, and humorous storytelling to get its message across.

This book is translated from its original German.

How To Rule the World:
Lessons in Conquest by the Modern Prince

by J.F. Cummings (2008)
ISBN: 978-4-902837-00-5

Following in the satirical tradition of Niccolo Machiavelli (*The Prince*) and Jonathon Swift (*Gulliver's Travels/A Modest Proposal*), *How to Rule the World* provides a commentary on today's "Modern World" and the "forces" that govern it. This is done in the voice of "civilization's" greatest supporter, an advisor to Prince. *How to Rule the World* is a modern adaptation of Machiavelli's *The Prince*. The author provides the reader, the Prince, with a methodology of non-invasive influence and control that will grant him sovereignty over his or her desired target nation-state and eventually over the world at large.

How To Rule the World shows the modern Prince how to utilize "Modern Ideals" such as free trade, democratic governance, human rights, freedom and individual rights, rule of law, and free press to exert control over other nations and convince them to collaborate in their own domination and exploitation through their quest to do whatever is required of them to be accepted as "Developed", "Modern" nations. Through adherence to the methodology of conquest explained the book the Prince will be granted access to the psyche of the target nation's population and will be able to redefine its very sense of worth and self-definition.

How to Rule the World, is written in first-person like *The Prince*, and is a conversation with the reader, leading to self-examination his or her own value system, thought processes, and concepts of human nature. It provides a forum through which the reader can determine his or her position in the world and within their own psyche as 'Prince' or 'subject', and how the actions of both impact on the very sustainability of the human species.

Ordering blue ocean press books:

Individual orders:

> Books can be purchased and ordered from your local bookstore.

> Books can also be purchased online through retailers such as: the Amazon.com family (amazon.com, amazon.co.jp, amazon.co.uk, amazon.fr, amazon.ca, amazon.de), Barnes and Nobles (bn.com, barnesandnobles.com), Powells.com, Abebooks.com, Alibris.com

Institutional Buyers, Booksellers, and Libraries:

> Books can be ordered from the following distributors and wholesalers:

> *U.S. and Canada*
> Ingram Book Group (ipage/Ingram, Ingram Library Services, Ingram International)
> Baker & Taylor
> NACSCORP (a wholly-owned, for-profit subsidiary of the National Association of College Stores)

> *U.K. and Rest of the World*
> Gardners Books
> Bertrams
> Baker & Taylor
> Ingram International